# Crossroads of the Southwest

# Crossroads of the Southwest: Culture, Identity, and Migration in Arizona's Safford Basin

Edited by

David E. Purcell

Cambridge Scholars Publishing

Crossroads of the Southwest: Culture, Identity, and Migration in Arizona's Safford Basin, Edited by David E. Purcell

This book first published 2008 by

Cambridge Scholars Publishing

15 Angerton Gardens, Newcastle, NE5 2JA, UK

British Library Cataloguing in Publication Data
A catalogue record for this book is available from the British Library



ISBN (10): 1-84718-480-4, ISBN (13): 9781847184801

# TABLE OF CONTENTS

# List of Illustrations

# LIST OF TABLES

# ACKNOWLEDGMENTS

The Safford Symposium was jointly presented by the Arizona Archaeological Council (AAC), Eastern Arizona College (EAC), and the Center for Desert Archaeology (CDA), with sponsorship by Carter + Burgess as the Fall 2005 meeting of the AAC. It was held at the Thatcher Campus of Eastern Arizona College on October 28-30, 2005. The meeting was organized by David E. Purcell, with the assistance of Jeff Altschul, Linda Blan, Mark Bryce, Jeff Clark, William Doelle, Pat Gilman, Sarah Herr, Jeff Homburg, Pat Lyons, Anna Neuzil, and M. Kyle Woodson.

The research on which Chapter Two was based was sponsored by the Arizona Department of Transportation (ADOT) and was conducted separately by Archaeological Research Services, Inc. (ARS), Tempe, Arizona, and Desert Archaeology, Inc. (DAI), Tucson, Arizona. Many people in each organization contributed to the technical reports of these projects, both in the field and during report production, but the direction, experience, and support of Lyle Stone (ARS) and William Doelle (DAI) was crucial to the success of these projects. The forbearance of Dawn and David Greenwald of Four Corners Research, Tularosa, New Mexico, is also acknowledged for the two years during which the U.S. 191 report draft was completed for previous employer ARS. Western Mapping, Tucson, Arizona, provided the maps of the U.S. 70 project and Shearon Vaughn of ARS drafted the artifact illustrations from Roadrunner Village. We are indebted to them for the use of these illustrations.

The following individuals have graciously provided their time, labor, and information to the betterment of Chapter Three: David R. Abbott; Jacque Attaway (Graham County Assessor's Office), Michael Bryce (Graham County General Services Office), Jeffery J. Clark, Verna Rae Colvin, John V. Cotter, Joseph S. Crary, Brooks Curtis, Myrna Curtis, Lee DeWester, William E. Doolittle, Paul R. Fish, Suzanne K. Fish, Jerry B. Howard, Gary A. Huckleberry, Robert C. Hunt, Annick Lascaux, Jonathan B. Mabry, Anna A. Neuzil, David E. Purcell, Anna Rago, Ann F. Ramenofsky, Staff members of the Gila Valley Irrigation District Office in Safford, Staff members (Colleen Edington, Betty Jo Ferrin, and Mel Jones) of the Graham County Historical Society, Kathy Valenzuela

(Graham County Records Office), Jennifer Varin (District Conservationist, Natural Resources Conservation Service, U.S.D.A.), Henry D. Wallace, David R. Wilcox, Samuel M. Wilson, and M. Kyle Woodson. It is with profound gratitude that we acknowledge their contributions.

The work described in Chapter Four was performed under a contract with Phelps Dodge Safford, Inc. The author thanks John Korolsky, of PD-Safford, as well as Tom Euler, Eric Petersen, Heather West, India Hesse, and Lara Mitchell of SWCA Environmental Consultants for their contributions to the success of the project.

Some of the results included in Chapter Five were obtained by Lone Mountain Archaeological Services, Inc., while under contract with Fort Bliss. Other data were gathered under volunteer agreements with the Bureau of Land Management, with special assistance provided by Anna Rago, and the Forest Service, through cooperation with Mary Farrell and William Gillespie.

# INTRODUCTION

# *THE SAFFORD SYMPOSIUM*: NEW RESEARCH IN THE OLD SOUTHWEST

DAVID E. PURCELL

During the fall of 2005, the organization of professional archaeologists in Arizona—the Arizona Archaeological Council—organized a conference on the campus of Eastern Arizona College in Thatcher, Arizona, to present the results of recent research in Thatcher and other locations in the Safford Basin (Figure One). The Center for Desert Archaeology, Eastern Arizona College, and the AAC presented the conference with sponsorship from Carter+Burgess, Inc. This book includes greatly revised versions of some of the presentations made at that conference.

Recent archaeological findings from the Southwestern United States, the subject of this book, is also the topic of hundreds of other titles currently in print. What more can be learned of the prehistory of this one region? Much, as the following chapters will demonstrate. The Southwest is a topographically and ecologically variable area of vast proportions, encompassing portions of the states of Arizona, New Mexico, Colorado, Utah, and Nevada, as well as the Mexican states of Sonora and Chihuahua. The vast distances of desert, sage-plain, and pine forest have been the home of many different peoples since the waning days of the last ice age, and perhaps before, as new findings from throughout the Americas push ever earlier the dates for peopling of the Americas. The level of intensity of research into the American past is uneven, however, even within the Southwest, where districts such as Mesa Verde, Chaco Canyon, and the Tucson vicinity are known in minute detail. The peripheral areas—the valley of the Colorado River or the edges of the Great Plains—are not so well described. This book is an introduction to the prehistory of another of those less-well-known areas, the Safford Basin of Arizona (Figure One).

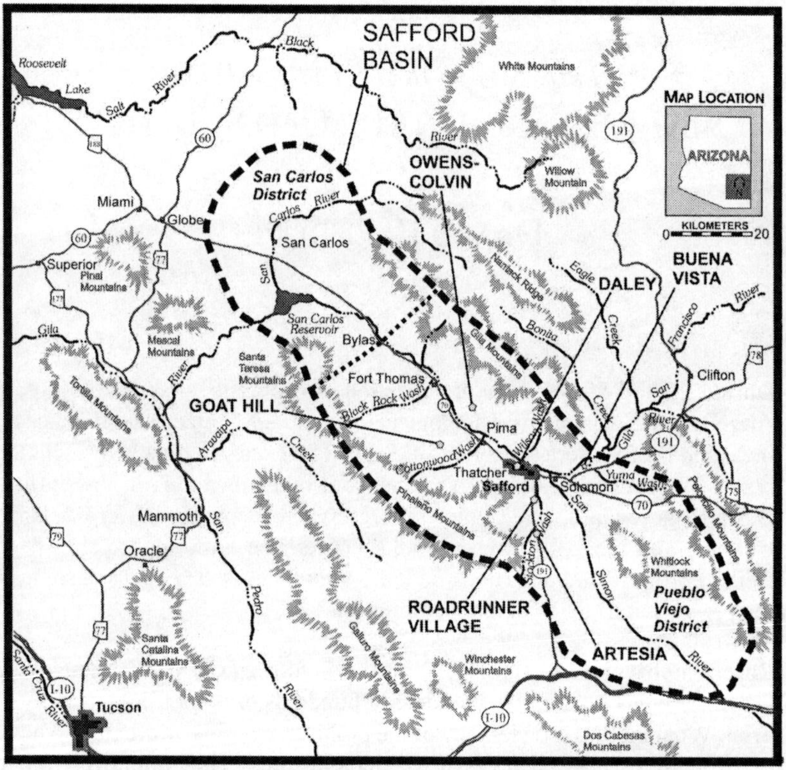


Figure 1: The Safford Basin, after Clark et al. 2004, showing the location of selected archaeological sites and districts discussed in this book. Some sites are not depicted to prevent unauthorized visitation (as required by the permits and other agreements the authors of this book hold with land managing agencies), while other sites are shown in greater detail—than is available at the scale of this map—in the pertinent chapters.

The Safford Basin is one of a series of topographical basins in the southeastern quadrant of Arizona. It encompasses part of the valley of the Gila River, one of the region's major waterways and a principal tributary of the Colorado River, draining southwestern New Mexico and southern Arizona. Gregory (1995:38-40), Doyel (2000), and Clark, Nials, and Vint

(2004) have defined the Safford Basin as that portion of the Upper Gila River Valley from the Sanchez Gorge (or Gila Box) downstream to the Mescal Mountains, including the *bajadas* (desert mountain slopes) of the Gila Mountains to the north, and the Santa Teresa and Pinaleño Mountains to the south. I include the valley of the San Simon River, as there appear to be no topographic or cultural barriers to its participation in cultural developments along the Gila River. The Safford Basin has been divided at the settlement of Bylas into two archaeological districts, the San Carlos and the Pueblo Viejo (Gregory 1995, cf. Clark et al. 2004). The cultural diversity encompassed by the Safford Basin may reflect the topographic and climatic variability of this area, in particular, the conjunction of the Sonoran Desert, Chihuahuan Desert, and Mogollon Highlands (Doyel 2000:Figure 13.1; Lekson 2000:Figure 12.1; Whittlesey et al. 2000). Each of these distinct environments describes particular associations of plants, animals, minerals, soils, and other resources, providing differing opportunities and obstacles for prehistoric peoples to live. Furthermore, since the earliest archaeological studies, it has been observed that during the Ceramic Period the Safford Basin includes cultural manifestations typical of all of the surrounding regions, groups that archaeologists call the Mogollon, Hohokam, Anasazi, and Salado. The Apache and their ancestors also made the Safford Basin their home, perhaps beginning in the late Ceramic Period.

The Safford Basin was an early focus of both meaningful historical observations of native peoples and archaeological ruins, as well as professional archaeological study. Written descriptions of archaeological sites and native peoples date to the Spanish Entrada of Coronado in 1540 (Hartmann and Flint 2001; cf. Chapter Five, this volume). The earliest professional archaeological observations were recorded in the 1880s (Bandelier 1892). Barely more than a century later, archaeologists have outlined the sequence and timing of major cultural developments (Figure Two), the attributes of prehistoric traditions from different districts, and the relationship between living native peoples and those of the prehistoric period. Through the decades, however, the attentions of professional archaeologists became increasingly focused on the Colorado Plateau, Rio Grande Valley, Mimbres Valley, Salt-Gila Basin, and Tucson Basin. The Safford Basin drifted to the margins of zones of archaeological interest (see Chapter Six). With the accumulation of data has come the need and desire to ask new questions about the past and revisit some old theories. Because much of the archaeological research conducted in the United States since the mid-1960s is contractual heritage resource management, rather than

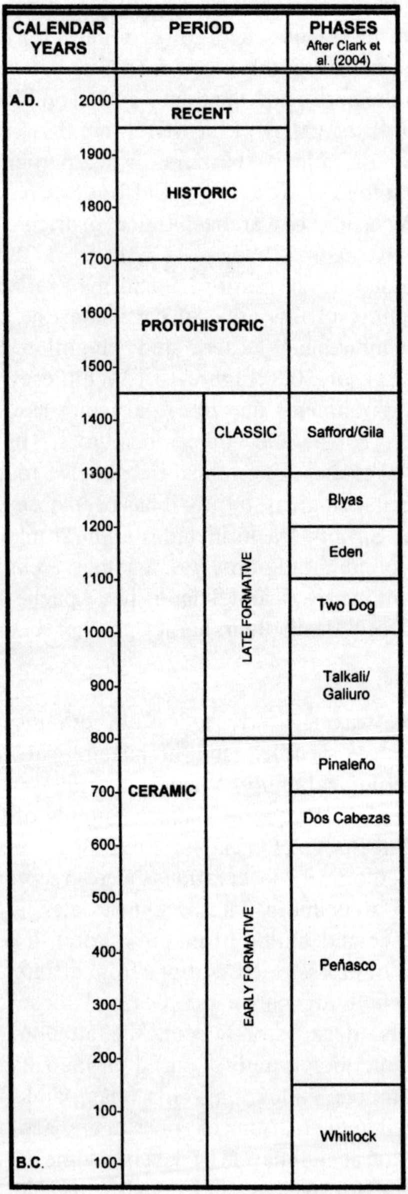


Figure 2: Post-Archaic period chronology of the Safford Basin, Arizona.

purely academic in nature, the regional focus of studies has gradually changed. It is growth and development of the land for homes, mines, farms, and highways that drives the new research now, and these activities sometimes fall in areas outside of traditional (i.e., preferred) regions of study. In the desert basin and adjacent mountains around Safford, Arizona, highway improvements, copper mines, and astronomical facilities have drawn a new generation of archaeologists. Academic archaeologists have also renewed their interest in the region, and are working with amateur and contractual archaeologists, as demonstrated by the authors of this volume.

The recent research presented at the Safford Symposium demonstrates that this desert basin contains such a diversity of archaeological sites that prehistorically it was at the crossroads of events in the Southwest, and is a natural laboratory to revisit some of the central themes of archaeology: migration, culture, and identity. Each of these topics has been part of archaeology in North America since its inception as a discipline, and before, as discussed below. This volume showcases the application of these theories to current, cutting-edge research in the Safford Basin.

## Migration

Migration is the physical relocation of an individual or species on a permanent basis. A seasonal change in residence is not a migration, but transhumance, an approach to climatic and landscape variability that has (and still) supports the economies of hunter-gatherer societies in many of the world's ecotones. In anthropology, migration is one means by which cultural information—beliefs, language, or manufacturing practices—is diffused from one region to another or one social group to another. Snow (1995:72) states that "One of the most enduring of archaeological problems has been the detection, confirmation, and explanation of human migration." In some instances, such as DiPeso's (1974) *pochteca* model, the migrants were individuals of a specialized class or position who journeyed north from Mexico to trade with groups in the Southwest, and in doing so, dispersed important cultural innovations, either intentionally or by local imitation of their example. In other instances, families, villages, or entire populations relocated to another region, a movement known formally as a diaspora but colloquially as a migration. It is the latter meaning encompassed by the research presented in this book.

In the hands of most processualists of the "New Archaeology" school, which dominated Southwestern archaeology from the 1960s to the 1990s, migration was set aside as an agent of social change (Hegmon 2003:227-228). Migration clashed with the dominant model of in-place cultural evolution (Jones and Klar 2005:457-458) and was often explained away as a ". . . special and generally unprecedented consequence of the rapid displacements, population declines, and technological changes brought on by contact with Europeans" (Snow 1996:794). Diffusion, if it was discussed at all, came to be used almost as a black-box process by which cultural information was transmitted without the movement of people (Bettinger and Eerkens 1999:236), either through contact from down-the-line trade (DiPeso 1974) or other unspecified and archaeologically unobservable means (Ford 1969; but see Reid 1997a). The gradualism of the processualist approach is sharply at odds with some native oral traditions, such as those of the Hopi and O'odham (Teague 1993), and with the considerable archaeological evidence for frequent, large-scale movements of people within the Southwest (Bernardini 2005; Coffey 2006; Reid and Whittlesey 2007; Wilshusen and Ortman 1999) and to the Southwest from the Great Basin (Young and Bettinger 1992).

As an agent of social change, migration has enjoyed favor cyclically in anthropological and archaeological theory, and is once again in vogue, as papers in this book demonstrate. New data from the Safford Basin point convincingly to migration as a central agent of social change in the Southwest, but we must also acknowledge that the restoration of migration theory may benefit from generational opposition to old paradigms by younger scholars, or to the prevalence of immigration in contemporary political discussion, especially in Arizona. Arizona is the destination for millions of migrants from Mexico and Central America, but also for millions of internal migrants from California and the old industrial region of the Midwestern states. The resulting clash of cultures from the heartlands of the United States and Mexico is an everyday event to the residents of Arizona, and of national political concern. The study of migration has become of central importance to the governments of many industrialized nations (Anthony 1990), whose wealth draws millions of disadvantaged people; in turn, archaeologists have renewed their interest in the topic.

Contributors to this volume—many of whom are already recognized for their work on the issue of migration—explore the number of migrations, their origins and process, and the effects of these movements on the cultures and ethnic identity of the Safford Basin. Neuzil (2006a) has suggested that there have been three migrations to the Safford Basin, beginning in the eleventh or twelfth centuries with Mimbres Mogollon, then followed by two immigrations of Anasazi or Ancestral Pueblos from the Colorado Plateau. However, the appearance of Hohokam villages in the region may mark an even earlier movement of peoples upstream along the Gila River from ancestral villages along the Middle Gila. Neuzil and Woodson (Chapter One) provide a detailed discussion of current archaeological models for migration and an overview of its history in Southwestern archaeological theory. Purcell and Clark (Chapter Two) examine the relationships between local villages of the Safford Basin and their neighbors at two sites, leading to an examination of whose tradition is local, in a multi-cultural landscape. Seymour (Chapter Five) expands the field of inquiry beyond the agriculturalists of the Ceramic period to investigate the potential for early immigration of Athapaskan-speaking nomads, and the evidence for similar migrations to go unnoticed in the archaeological record due to differences between the migration strategies of sedentary and mobile populations.

## Culture

In the Southwest, culture is often confused or confounded with culture area. A culture area is a geographical region in which a suite of material traits is distributed: pottery, house architecture, grinding stones, and other durable remains (Gladwin and Gladwin 1929). In rare instances, the perishable component of material culture is also recovered, affording a more complete view of the items made in a given area at a given time. Thus, the Hohokam culture area of southern Arizona is conceptualized as the homeland of people ("the Hohokam") practicing the Hohokam culture (Haury 1976), as opposed to the Anasazi or ancestral Pueblo culture of the Colorado Plateau (Kidder 1975), Mogollon culture of the Arizona-New Mexico highlands (Haury 1936; Wheat 1955), or the Patayan culture of the Lower Colorado valley (Colton 1939; Rogers 1945).

Culture, as an overarching system of beliefs and practices (operating at both the conscious "ascribed" level and the sub-conscious or "learned" level) is believed to direct the manufacture of the goods and buildings found in the archaeological record. The practice of a culture in a given region thus leads to the observed distribution of the material culture studied by archaeologists: in the cultural ecology approach of the processualist's model, cultures come into existence in response to major climatic events, and depart under similar circumstances, with the intervening centuries divided on the basis of minor changes to the basic program into "phases" (e.g. Berry 1982, cf. Irwin-Williams 1985:45). Perhaps the most obvious of many problems with this approach is its environmental determinism and lack of human agency. Only natural, climatological events are allowed as agents of historic change, with humans passively reacting and "adapting" themselves to the new circumstances.

The utility of the culture area approach in organizing the bewildering array of materials being observed and collected during the early decades of archaeological research in the Southwest cannot be denied, but it clearly hinders a more fine-grained or nuanced approach to understanding the interplay of cultural practices and specific, historical events in a given location (Bernardini 2005; Hegmon 2003:224). This is particularly true if that location is peripheral to the "great" culture areas, and the residents of the area inconveniently mix traits from different cultures. Regardless of well-reasoned criticisms to the contrary (Bernardini 2005:31; Cameron 2005; but acknowledging the veracity of their arguments), culture and culture areas remain basic units for analyzing and describing the past interactions of the people in the Southwest. Legal requirements to assign a cultural affiliation to sites and human remains underscore the utility of culture as a convenient shorthand for lists of material traits, themselves an abbreviation of the human behavior and human hands that produced them.

Our examination of culture in the Safford Basin takes many forms, but is unified by the awareness that different settlement patterns, economic systems, and traditions of learning how to make the elements of material culture—pottery, grinding stones, cutting tools, or houses—are represented in the Safford Basin diachronically and synchronically: through time and at the same time. Adjacent villages may exhibit very different structure in house locations, house construction, and associated artifacts, although they appear to have been occupied at the same time. Even portions of single settlements or houses within villages exhibit a level of dissimilarity that sets the Safford Basin apart from other regions. From the beginning of archaeological studies in the Safford Basin, these differences have been explained by the immigration of groups from the Hohokam heartland to the west, Mogollon regions to the north and east, and the Anasazi lands of the north, and in the subsequent blending of material culture traits with the San Simon Mogollon, who are generally viewed as an indigenous agricultural population that developed directly from earlier hunter-gatherer populations of the region (Sayles 1945).

To varying degrees, all of the chapters that follow explicitly evaluate the Safford Basin as a place that simultaneously hosted people of different cultures and what that means in the effect of migration on archaeological cultures. The recognition of sites that appear to have been occupied by people of different cultures in the Safford Basin is often the basis for the interpretation that such a pattern results from migration. Neuzil and Woodson (Chapter One) track the arrival of immigrants from the Colorado Plateau through changes in ceramic and architectural traditions. Similarly, Seymour (Chapter Five) identifies the unique artifacts and features that are diagnostic of the late prehistoric migration of ancestral Apache to the Safford region. Purcell and Clark (Chapter Two) look at the relationships between presumed "local" cultures and "immigrant" cultures through multiple traits, and the implications for a greater time-depth to ethnogenesis in the Safford Basin than has been previously modeled. By including nearby prehistoric and historic-age archaeological sites in their study of canals in the Safford Basin, Neely and Murphy (Chapter Three) re-examine assumptions about which past cultures have constructed irrigations systems, and when. Ezzo (Chapter Four) utilizes the motifs of rock art to examine the occasional and ephemeral uses of the upland landscape of the Safford Basin, whereas the place of the Safford Basin in the big picture of Southwestern polities is explored by Lekson (Chapter Six).

## Identity

The concept of identity seems central to social and political life in the post-industrial world, yet it remains one of the most difficult aspects of human culture to recover from archaeological contexts, in part, because the very definition of identity seems both elusive and non-exclusive with other concepts such as culture. Indeed, as noted above, the very concept of archaeological culture has been challenged on the basis of the obvious disparity between self-reported identity and identity as described by material culture (Cameron 2005). Bernardini (2005:49) succinctly notes that "The primary lesson of [his study] is that prehistoric identity resided in social groups, not geographic territories." The ethnohistoric Pueblos stand as a daunting cautionary tale to archaeologists, exhibiting similar material culture but inhabited by peoples of radically different languages, and identities. The social distinctions that most archaeologists study most of the time operated at the subconscious level, the ways of making pots and grinding stones and houses that are learned at an early age and are applied through life. These patterns of material culture tend to form large spatial distributions that are identified as archaeological cultures. The practitioners of these cultures may speak different dialects or even different languages, and at the highest levels of identity may recognize a myriad of different tribes, nations, clans, moieties, sodalities, or other social statuses among their number: "Prehistoric cultures are identified through physical evidence, while historic populations have been identified principally in terms of linguistic identifications" (Teague 1993). Perhaps the critics of archaeological cultures should acknowledge that identity and culture are not synonyms and that human social life exists at multiple levels, some very much at the forefront of daily social existence (which equate roughly with the "emic" or self-reported structure of societies) and others existing at a subconscious level ("etic" or identified through external or non-participant observation, as by anthropologists). Unfortunately for archaeologists, most of culture is information that is probably not recoverable from the archaeological record, which holds only the most durable of goods such as stone, pottery, and wood, but rarely fully articulated ideas. And identity, in the ethnic sense, may not be identifiable through osteological analysis of human remains, or even DNA. Archaeology of the historic or written period often reveals significant discrepancies between written accounts of an event, and the physical evidence of that event: "As in many other cases, an archaeologist's systematic, materialist view of a society contrasts with the written record of the inhabitant's self-perceptions" (Rathje et al. 1992:442). This phenomenon has been documented in the discrepancy between self-reported consumption of alcoholic beverages by residents of Tucson, Arizona, and the material evidence for much greater consumption as revealed by examination of their garbage (Rathje et al. 1992).

Identity, then, is a more sharply focused and immediate idea than culture. For our purposes, culture is the archaeologically derived and defined material culture of a specific area and time, combining the culture area concept described above with a particular phase of development. Bernardini (2005:34-35) has advocated for such a time/space approach in which culture is viewed as an essential impermanent construction, always in the process of becoming something else, even as we examine it. Thus, an archaeological residential site of the Salt River Valley in central Arizona is demonstrably different at A.D. 1000 from a site of similar size on Mesa Verde in Colorado, along the Mimbres River in New Mexico, or near Yuma, Arizona, but is also different from an archaeological residential site along the Salt River one generation later. All such sites have in common the use of pottery vessels, a maize-dependant diet, and the use of grinding stones to prepare maize, commonalities that define the Southwestern culture area, which is different on these bases from similar broad trends of the Great Plains and Great Basin, adjacent North American culture areas. In our examination of the Safford Basin, identity will then be applied to the higher, conscious levels of culture, levels that may have some correspondence with how those people viewed themselves. In this instance, it may be viewed archaeologically as an individual's preference to retain the practice of pottery making as learned from his or her parents, but applied to new materials and styles of decoration. The Maverick Mountain type of pottery is made using some techniques typical of the lower deserts, but with designs and layout of the Anasazi (Lyons and Lindsay 2006). At the Anasazi immigrant enclave of Goat Hill, the entire settlement was built in the style of the homeland (Woodson 1995, 1999).

Identity is assumed, but not explicitly discussed in many of the papers contributed to this book, with the exception of Neuzil and Woodson (Chapter One), who provide a detailed discussion of identity in support of their argument. Ezzo attempts to examine identity through rock art in Chapter Four, in assessing who produced the designs found along the north side of the Safford Basin. And Seymour (Chapter Five), in addressing the immigration of Athapaskan speakers to the southern Southwest, assesses changes to their material culture that bring them from near invisibility in the archaeological and historical record to the very distinctive and multiple identities they exhibited by the 1600s.

As will be shown by the papers contributed to this volume, the concepts of migration, culture, and identity are not discrete or abstract, but inherent in the complex and dynamic interactions of people in places like Arizona's Safford Basin. Beyond the dry associations of trash and soil strata that we archaeologists study are people and their behavior. Through the various studies described in this book we can glimpse some of the people who lived in the Safford Basin in prehistory, both those who were born there and those who came to live there from some other place. In this aspect, the Old Southwest seems very similar to the New Southwest, dynamic, multi-cultural, and multi-ethnic. Much remains to be done to further examine the interplay of migration, culture, and identity in the region. With hope, this volume represents the beginning of a new era of research.

# CHAPTER ONE

# MIGRATION AND IDENTITY IN THE SAFFORD AND ARAVAIPA VALLEYS OF SOUTHEASTERN ARIZONA

ANNA A. NEUZIL AND M. KYLE WOODSON

## Introduction

Significant research on migration and identity in the past has been conducted recently, greatly advancing how archaeologists and others in the social sciences approach these topics (Barth 1994; Bentley 1987; Birkeland 2000; Brettell and Hollifield 2000; Clark 2001; Craib 1998; Cuba and Hummon 1993; Díaz-Andreu 1996; Frayne and Pendelton 2001; Gardner 2002; Haug 2002; Herr 2001; Jenkins 1996; Lekson et al. 2002; Levy and Holl 2002; Lyons 2003; Marshall and Foster 2002; Meskell 2001, 2002; Mills 1998; Schnapper 1994; Stone 2003; Upton 1996; Vertovec 2001). Theory and method have both become more rigorous, so that migration can now be more fruitfully and meaningfully examined. At the same time, new research on identity in the past has expanded our understanding of the distinctions between individual and group or social identity, and how they relate to each other. Using the example of a wave of migration into the Safford and Aravaipa areas of southeastern Arizona in the late thirteenth and fourteenth centuries, we will examine the social consequences of migration and its impact on the expression of identity.

## Migration

Migration is best defined as

> . . . a long-term residential relocation beyond community boundaries by one or more discrete social units as the result of a perceived decrease in the benefits of remaining residentially stable or a perceived increase in the benefits of relocating to prospective destinations (Clark 2001:2).

Since the resurgence of archaeological interest in migration research inspired by Anthony (1990), significant advances have been made in identifying occurrences of migration, and understanding why migration occurs, how people migrate, and the social and economic consequences of migration. Underlying this research is a conception of migration as a patterned and structured process with recognizable and understandable causes and consequences, both in the past and present (Anthony 1990). Because migration was an almost taboo topic in American archaeology during the 1960s, 1970s, and 1980s, very little research on migration was conducted during this time (Anthony 1990, 1994, 1997). As a result, contemporary archaeological inquiry into migration is informed by research on migration in a variety of other disciplines (Brettell and Hollifield 2000).

Recent research has demonstrated that voluntary migration occurs at a variety of scales, each having a different impact dependent on the size and composition of migrant groups and the level of interaction between migrant and indigenous populations. Social scales of migration range from the individual, to the household, to the suprahousehold group, to the community (see Table One). Household level migration is generally the smallest scale of migration that can be recognized in the archaeological record through material culture that reflects shared kinship and settlement history (Bernardini 2002; Mills 1998; Neuzil 2001; Van Keuren 1999). Suprahousehold level migration, also called a "site-unit intrusion," consists of the movement of multiple households organized as a larger social unit (Wauchope 1956; Woodson 1995, 1999). This is well represented by the classic example at Point of Pines defined by Haury (1958; see also Di Peso 1958), and can be recognized by the appearance of non-local material culture such as ceramics, architecture, and burial practices that contrast markedly with local material culture and practices and can be traced to a known homeland. Community level is the largest scale of migration, representing the movement of multiple households and suprahousehold groups into an area, and can also be recognized by the appearance of non-local material culture combined with a substantial increase in population. All three scales of migration are recognized by the appearance of non-local material culture and an increase in population. Variability in the relative abundance of non-local material culture and the scale of population increase distinguish among the three levels of migration; in other words, intrusive material culture is more apparent and population increases are more pronounced for community level migration than for suprahousehold or household level migration.

## Why Do People Migrate?

As is evident from Clark's (2001) definition, migration results from changes in the social and economic benefits of remaining in place in contrast to moving to a new location. Migration involves risk, which can be reduced by interaction between populations in the source and destination through kin relations, information flow, and previous migration experiences (Bernardini 2005). The reduction of risk further increases the probability of movement in response to declining living conditions (Anthony 1990).

Motives for migration are often framed in terms of push factors from the homeland and pull factors into the destination (Anthony 1990; Herr and Clark 1997; Jett 1964; Lipe 1995; Schwartz 1970). Economics are often cited as the main reason for contemporary migrations (Frayne and Pendelton 2001; Heleniak 2004). In instances of migration in small-scale society contexts in the ancient past, changes in the environment are most often cited as push and pull factors. For example, Ahlstrom et al. (1995) and Lipe (1995) cite changes in precipitation, agricultural potential, water supply, erosion, and average temperature as the main reasons for the abandonment of the Mesa Verde region in the late thirteenth century. Similar environmental triggers are cited for migration from the Kayenta area around the same time (Dean 1996; Dean et al. 1994; Van West and Dean 2000), despite the fact that other research has demonstrated that the environment could have supported the existing population (with drastic changes in social organization) at the time people chose to move south (Kohler et al. 2005).

**Table One: Scale of Migration**

| Scale of Migration | Destination Area | Archaeological Signature | Examples | References |
|---|---|---|---|---|
| <u>Household</u> | Occupied | New houses or enclave(s) established within an existing settlement in a peripheral location<br>Appearance of subtle differences in domestic architecture (construction and room size), and ceramics (decoration style, technology), sometimes focused within one area of a settlement<br>Rapid construction of domestic architecture | Bailey Ruin<br>Grasshopper Pueblo<br>Gibbon Springs Site | Mills (1998)<br>Reid (1997)<br>Slaughter and Roberts, eds. (1996) |
| | Unoccupied | Single settlement established in a previously unoccupied area<br>Population increase<br>Widespread exchange networks reflecting areas with preexisting social contacts | Silver Creek Great Kiva sites | Herr (2001) |
| <u>Suprahouse-hold</u> "site-unit intrusion" | Occupied | Establishment of new settlement, enclaves found in a peripheral or defensible location<br>Distinctive and non-local material culture associated with new settlement – ceramic design style and technology, ritual and domestic architecture, food remains, lithic technology, burial practices, and architectural features, traceable to a homeland | Point of Pines<br>Goat Hill<br>Reeve Ruin<br>AZ CC:2:23(BLM) | Haury (1958)<br>Woodson (1999)<br>Di Peso (1958)<br>Neuzil (2005b) |
| | Unoccupied | --- | --- | --- |
| <u>Community</u> (multiple settlements) | Occupied | Multiple large settlements established suddenly in a previously occupied area, enclaves found in a peripheral and defensible location<br>Significant differences in domestic technology from the indigenous population (ceramics, domestic architecture, site layout) traceable to a homeland | Griffin Wash Locus A | Clark (2001) |
| | Unoccupied | Substantial population influx<br>Multiple socially linked large settlements established around the same time and general location in a previously unoccupied area<br>Distinct material culture traceable to a homeland | Homol'ovi sites | Adams (1998) |

Migration is ultimately a social strategy developed within the context of a particular culture. Social factors, perhaps influenced by the environment, likely played an important role in most migrations (Anthony 1994). Recent research on contemporary and historic migrations supports this assertion (Comas d'Argemir and Pujadas 1999; Herr and Clark 1997; Marshall and Foster 2002). Understanding the social motivations behind migrations is crucial, as some researchers have argued that framing the decision to migrate in terms of push and pull factors oversimplifies the process and ignores the agency of the migrating individuals and groups, which is key to understanding the process in its totality (Jones 1999; Marshall and Foster 2002).

## How Do People Migrate?

As a patterned process, migration has a number of recurrent elements in prehistoric, historic, and contemporary contexts. The process of migration to previously known areas differs dramatically from the colonization of new and completely unknown lands or frontiers that are uninhabited or sparsely inhabited (Rockman and Steele 2003). Migrants to empty landscapes or frontiers must deal with an entirely different suite of challenges than migrants who settle in areas that are known and inhabited, and models dealing with migration to empty landscapes are most applicable for understanding the movements of early small-scale forager groups (Herr 2001). Migrants that move into already occupied territory either move into unoccupied niches or settle alongside firstcomers and original inhabitants of the area as in a diaspora (Lilley 2004; see Mills 2007 for an in depth discussion of this division).

Migration into known areas first requires a flow of information between the homeland and the destination, which can be developed by long distance trade, exogamy, or shared resource procurement (Anthony 1990; Cameron 1995; Marshall and Foster 2002). This flow of information helps migrant groups make informed decisions about the benefits of potential destinations. Although initial information flow to migrant groups is often along kin lines, migrant groups are not always kin-based (Cameron 1995). However, social networks are crucial to the survival of migrants both during the process of migration and after arrival at the destination (Bernardini 2005; Haug 2002).

Once the destination has been chosen, migration itself can occur in a number of different ways. Anthony (1990:902-904; 1994) sets out two models of long distance migration seen in the archaeological record. In the first model, alternately called leapfrogging or chain migration, scouts bypass large areas while collecting information on resource potentials to relay back to the homeland, creating islands of settlement separated by significant expanses of vacant land. The second, the migration stream model, posits that migrants "proceed along well-defined routes toward specific destinations" (Anthony 1990:903), resulting in recognizable artifact distributions following a specific route. In this case, migration routes can be multidirectional as well as linear (Bernardini 2002; Reid 1997a). Current research on migration in the Southwest suggests that prehispanic migrations are best characterized by a combination of these two models, where migrants had specific destinations, and left routes traceable through artifacts, but also bypassed large areas and created islands of settlement (Haury 1958; Mills 1998; Lyons 2003). This is likely due, at least in part, to the unequal distribution of areas suitable for habitation and agriculture throughout the Greater Southwest.

## What are the social consequences of migration?

Inevitably, migration results in changes in both migrant and indigenous groups as new social situations between culturally and structurally distinct groups arise. The scope of these changes is dependent on the relative size and organization of the indigenous and migrant populations, the scale and organization of migration, previous relations between migrant and indigenous groups, and the level of integration between migrant and indigenous groups following migration. Substantial economic and social power inequalities may develop between migrant and indigenous groups following migration (Burmeister 2000), and this imbalance will necessitate new social institutions to integrate these disparate groups thereby increasing social complexity, or lead to instability, conflict, and possibly further migration (Bernardini 1998; Herr and Clark 1997; Lekson and Cameron 1995). A feeling of belonging or a sense of place can help to mitigate the desire to move again or to return to their homeland, although this takes time to develop (Frayne and Pendelton 2001; Haug 2002; Heleniak 2004; Jones 1999). The platform mounds of the Classic Period Tonto Basin and the ritual architecture of the Silver Creek area during the Pueblo III to Pueblo IV period transition are evidence of the integration of migrant and indigenous populations into a new social structure (Clark 2001; Mills 1998).

A part of the integration process may be a renegotiation of identity and sense of place for both groups (Marshall and Foster 2002; Schnapper 1994; Torres-Saillant 1997). In a broad sense, the identity of both migrant and indigenous groups may be renegotiated, persist, or become subsumed within others (Anthony 1997). Migrants are not likely to retain a distinct identity for an extended interval (Lucassen and Lucassen 1997), but may retain conservative elements of their pre-migration cultural background, while renegotiating their identity to fit in to their new social space (Levy and Holl 2002; Mills 2004). Such renegotiations of identity can lead to a feeling of belonging in two places, the homeland and the destination of migration, and can lead to multiple definitions of migrant identity, common among modern transnational migrants (Vertovec 2001).

## Identity

Despite, or perhaps as a result of its importance, identity has become a somewhat problematic concept in anthropological and archaeological literature. Researchers variably use the terms identity, social identity, ethnic identity, and ethnicity to refer to broadly similar concepts, often assuming that they can be used interchangeably (with some notable exceptions, see Jenkins 1996). For the purposes of this paper, identity will be defined following Jenkins (1996:4), as "the ways in which individuals and collectivities are distinguished in their social relations with other individuals and collectivities." These "ways of distinguishing" are operationalized and embedded in the social practices that constitute daily life (Jenkins 1996).

The concept of identity connotes both similarity and difference, often operationalized as an internal/external dichotomy: individual identity emphasizes differences between each individual, and social identity emphasizes similarities between individuals within a group and differences between groups (Jenkins 1996; Jones 1999). Identity is also expressed on a number of overlapping levels, depending on the situation and the audience. For example, a single person can express their identity based on their gender, family group, household membership, ethnic background, country or place of origin, occupation, beliefs, and a number of other levels. Some of these expressions of identity are more mutable than others. Ultimately all identities are social, in that they require another individual or group to exist and are constructed at social boundaries (Barth 1969a; Jenkins 1996). Furthermore, differences in social boundaries result in the situational construction and expression of identity. Interpretations of expressions of identity are similarly fluid among groups (Meskell 2002).

Identity is necessary for individuals to understand who they are in relation to those with whom they interact (Jenkins 1996). This is particularly the case in social situations where notions of identity are challenged, such as periods following migration (Jenkins 1996; Meskell 2002). Thus, reconstructing identity and its material expression can lead to greater understanding of the effects of new social situations that lead to unintended consequences, such as migration.

## Modeling the Consequences of Migration on Identity

Just as the archaeological signatures of different scales of migration vary, so do their impacts, depending on the relative size of migrant and indigenous populations and the social means available for integration. Table Two presents a model of the impacts of migrations to an occupied area, such as the Safford and Aravaipa valleys, on the expression of identity. While changes in the expression of identity are most likely to have been the result of a disparity in size between migrants and indigenous groups, other factors may also have played a role, such as the availability of social methods to integrate outsiders, the historical tolerance for outsiders due to earlier migrations, and the social and physical distance between groups. Migrants are usually at a disadvantage as a result of their newcomer status (Clark 2001). As this model makes apparent, migration can have a number of impacts on the identities of migrant and indigenous populations. Identities can be maintained as distinct from one another, renegotiated into a new identity, or subsumed into one another such that one population takes on the identity of the other.

**Table 2: Scale of Migration Model.**

| Scale | Size of Indigenous Population | Distance Traveled by Migrants | Length of Occupation Following Migration | Consequences for Identity |
|---|---|---|---|---|
| Household | Smaller | — | — | — |
| | Larger | Short | less than one generation | Identities are maintained |
| | | | more than one generation | Identities are renegotiated |
| | | Long | less than one generation | Migrants adopt indigenous identity |
| | | | more than one generation | Identities are renegotiated |
| Suprahousehold or Community | Smaller | Short | less than one generation | Identities are maintained |
| | | | more than one generation | Indigenous population adopts migrant identity |
| | | Long | less than one generation | Identities are maintained |
| | | | more than one generation | Indigenous population adopts migrant identity |
| | Larger | Short | less than one generation | Identities are maintained |
| | | | more than one generation | Identities are renegotiated |
| | | Long | less than one generation | Identities are maintained |
| | | | more than one generation | Identities are renegotiated |

## Migration and Identity in the Safford and Aravaipa Valleys of Southeastern Arizona

Despite the general paucity of research carried out in the Safford and Aravaipa valleys, migration and the arrival of migrants from northeastern Arizona during the late thirteenth and fourteenth centuries has been a recurring theme of research in these areas (Brown 1973; Hall and Clark 2004; Lindsay 1987; Wasley 1962; Welch 1995; Woodson 1995, 1999). Early research here looked toward a specific constellation of material traits in the archaeological record to identify potential migrant sites. However, very little of this research has systematically addressed and demonstrated where migrants came from, where they settled upon arrival, and how they interacted (or not) with the local population when they arrived. Recent research by the authors has addressed these questions and begun to fill in the gaps in our understanding of migration in the Safford and Aravaipa areas.

In 1994, Kyle Woodson (1995, 1999) led an excavation at the Goat Hill site (AZ CC:1:28[ASM])[1] to test the hypothesis that it was occupied by migrants. This masonry pueblo is located on top of a steep butte within Lefthand Canyon in the northern foothills of the Pinaleño Mountains. The site consists of a circular arrangement of 35 rooms surrounding a plaza with a D-shaped kiva that was occupied from about A.D. 1275 to 1315 (see Figure Three). Fieldwork included the excavation of 11 rooms and the kiva. Woodson concluded the site was occupied by migrants with close ties to the Western Anasazi (or Ancestral Pueblo) region in northeastern Arizona. This assessment was based on the occurrence of architectural styles that are common in sites from the Tusayan and Kayenta branches of the Western Anasazi region, including the D-shaped kiva, clusters or suites of 2 or 3 rooms, and entryboxes. Also, the decorated ceramic assemblage was dominated by Maverick Mountain Polychrome and Black-on-red, pottery types that exhibit designs with a stylistic grammar generally thought to have been developed by the Western Anasazi of the Colorado Plateau (Lindsay 1987; Lyons et al. 2005). These traits, along with the defensible hilltop location, stand in stark contrast to local Bylas phase settlement patterns and further highlight how the Goat Hill occupants were "outsiders."

Woodson's research at the Goat Hill site conclusively demonstrated that migrants from the Kayenta and Tusayan areas of northeastern Arizona settled in the Safford Valley in the late thirteenth century. Recent research by Anna Neuzil has suggested that at least two other sites in the valley, AZ

CC:2:23(BLM) (see Figure Four) and the Marijilda site (AZ CC:5:6[ASM]; see Figure Five) may have also been enclaves occupied exclusively by migrants, based largely on their decorated ceramic assemblages and domestic architectural traits. Furthermore, AZ CC:2:23(BLM) may have also been part of a larger group of enclave sites including AZ CC:2:33(BLM) (see Figure Six), and the Fischer site (AZ CC:2:100[ASM]; see Figure Seven). Recent investigations by Woodson at the Smith Tank site (AZ CC:1:133[ASM]; see Figure Eight) show it may have also been a migrant enclave potentially associated with other enclave sites in the vicinity. However, several lines of evidence from other sites that post-date A.D. 1275 throughout both the Safford and Aravaipa valleys suggest a combination of migrant and indigenous groups at these sites. This is largely based on data from ceramic assemblages and domestic architecture, collected by Neuzil from 2003 to 2004 for her dissertation research.

In the interval A.D. 1275-1325, ceramic assemblages typically exhibit greater diversity in ceramic types than sites occupied both before and after this time. Three lines of evidence suggest this diversity is largely due to the variety in social groups, including migrants, present at these sites. First, corrugated ceramics were analyzed in order to measure and examine their technological style, or the way in which they were formed and manufactured. Previous research has demonstrated that differences in technological style of corrugated ceramics can be indicative of discrete households with distinct enculturative backgrounds, as would be the case if migrant and indigenous groups lived at the same settlements (Neuzil 2001, 2005a). At many of the Classic Period sites in the Safford and Aravaipa valleys, particularly those that were relatively large (containing at least 50 rooms) differences in the technological style of corrugated ceramics were detected, suggesting several distinct households lived at these sites (Neuzil 2005b).

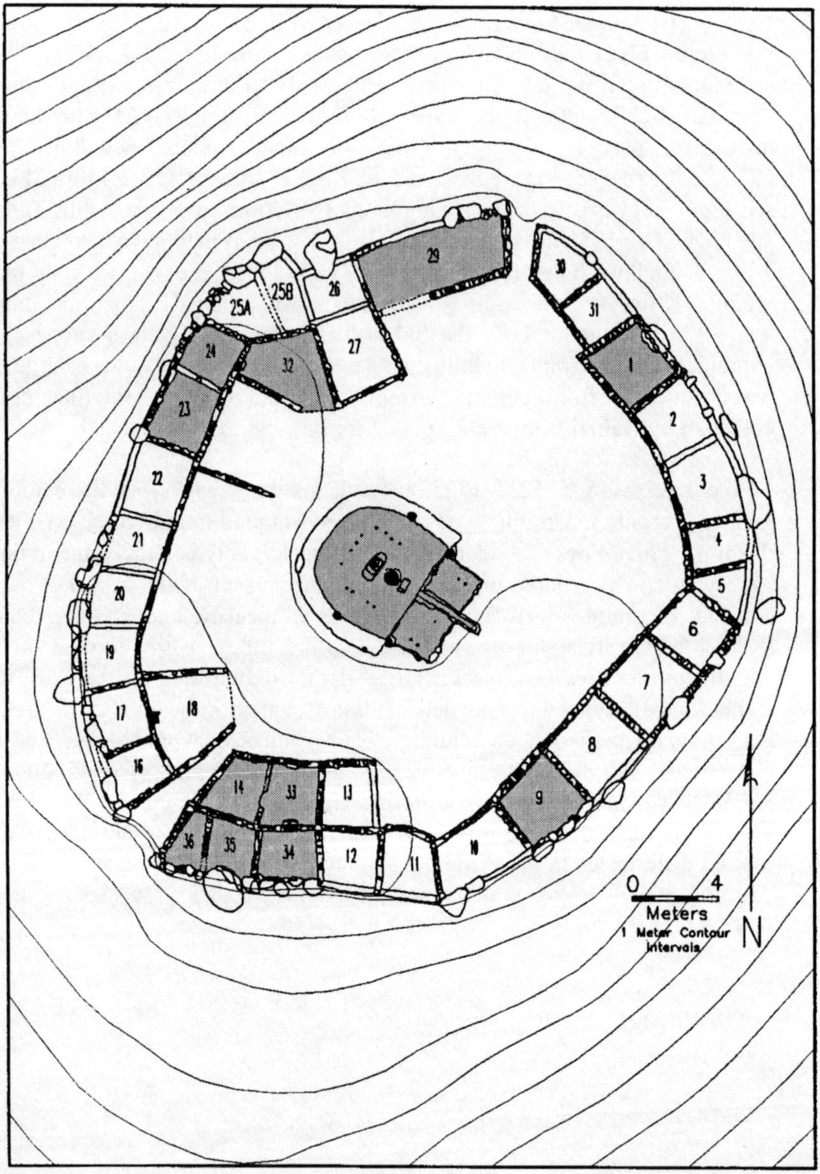


Figure 3: The Goat Hill site (AZ CC:1:28[ASM]; shaded rooms were excavated).

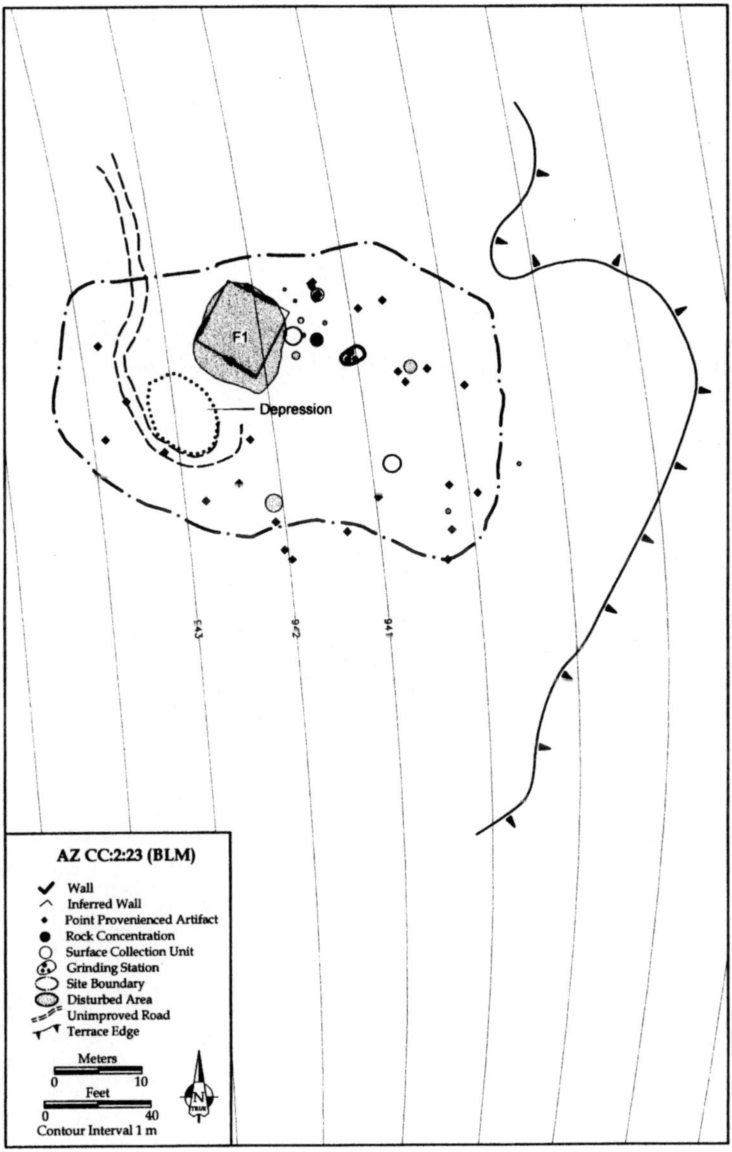


Figure 4: AZ CC:2:23(BLM).

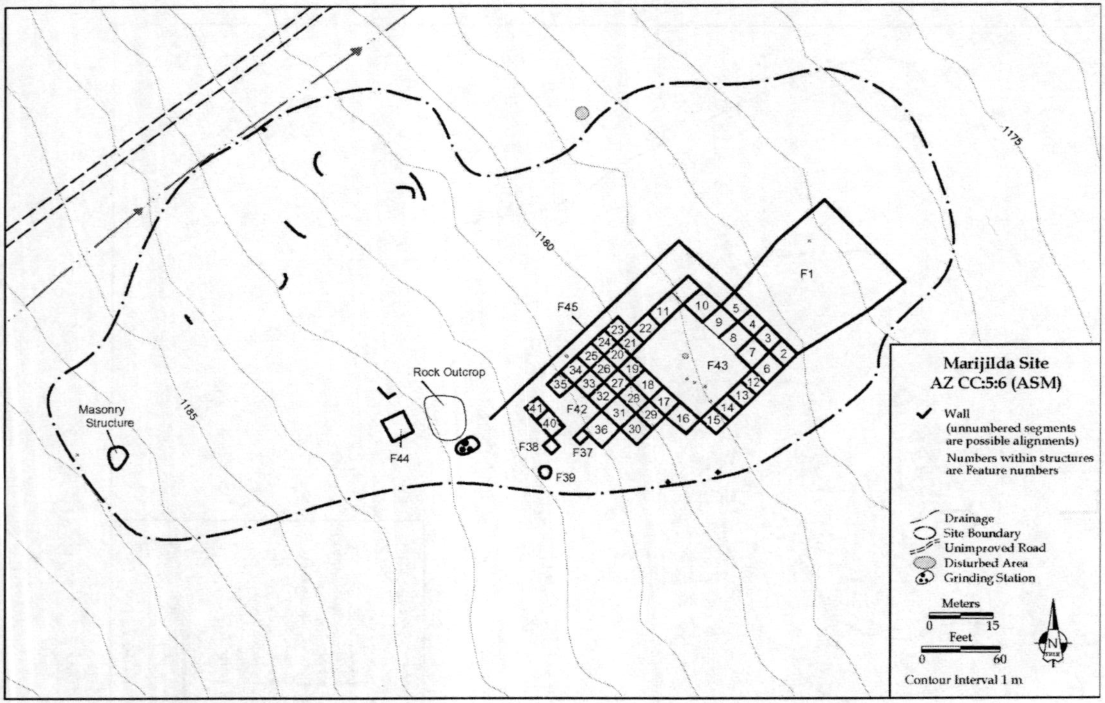


Figure 5: The Marijilda site, AZ CC:5:6(ASM)

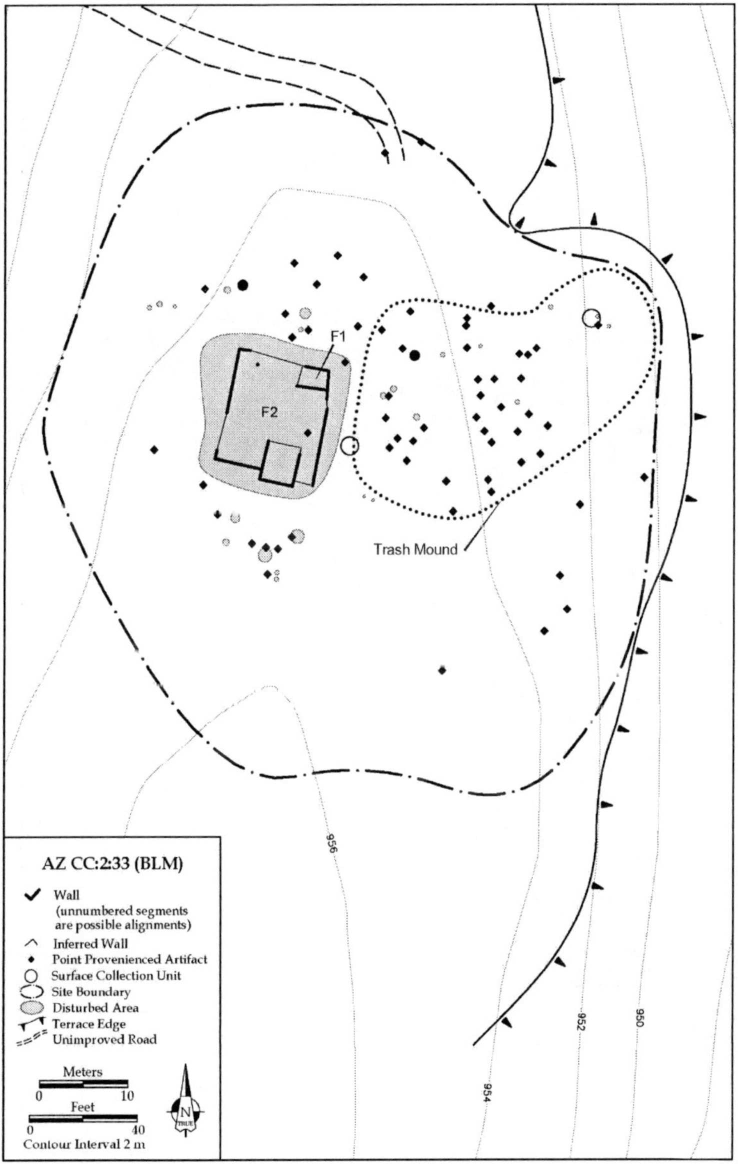


Figure 6: AZ CC:2:33(BLM)

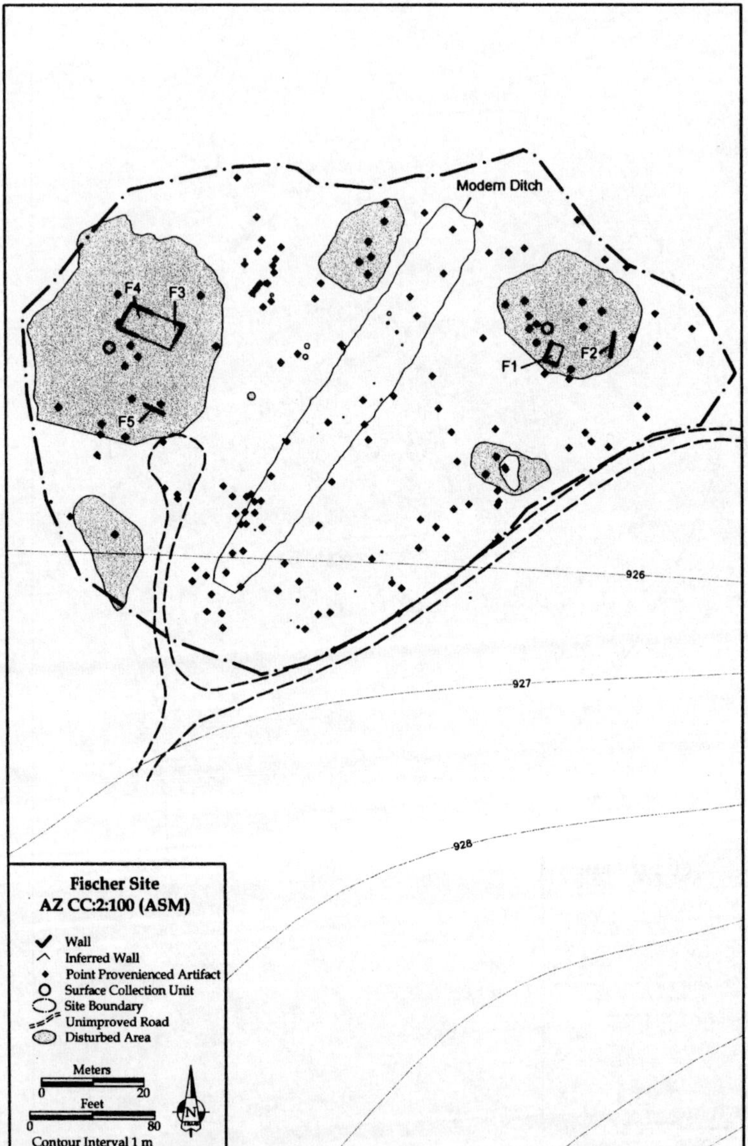


Figure 7: The Fischer site, AZ CC:2:100(ASM).

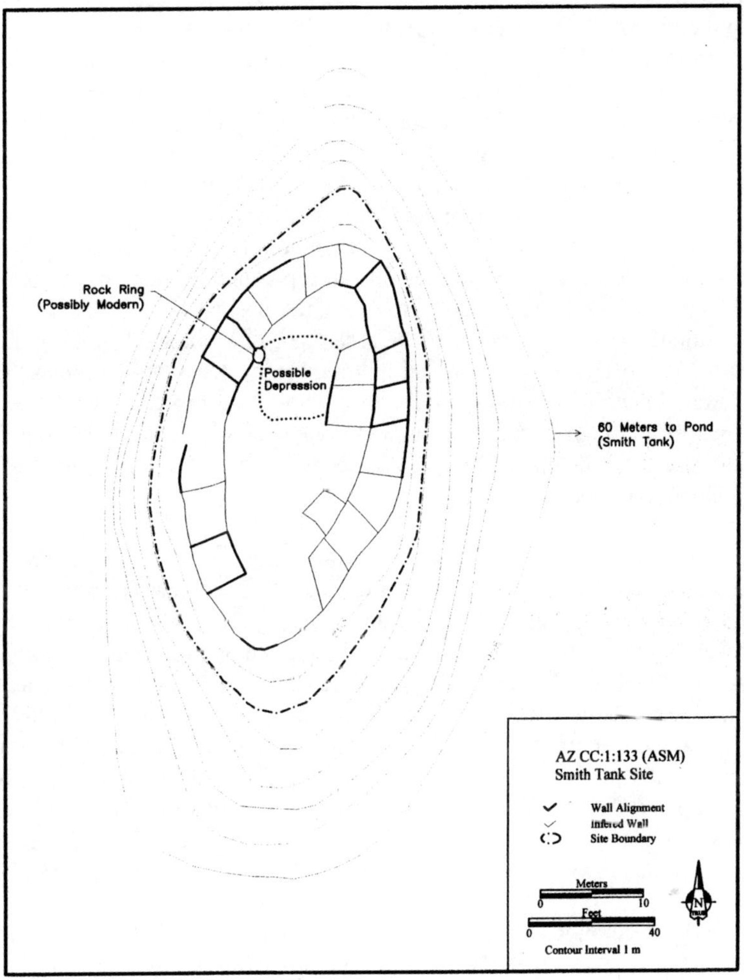


Figure 8: The Smith Tank site, AZ CC:1:133(ASM)

Taken alone, these data do not necessarily indicate the presence of migrant populations at Safford and Aravaipa valley sites. However, data from decorated ceramic assemblages provides further clues that the distinct social groups seen in the corrugated ceramic assemblages were actually composed of both migrants and indigenous groups. The decorated assemblages at all sites postdating A.D. 1275 contained both migrant associated ceramics, particularly Maverick Mountain Series, as well as those thought to have a local provenance, such as San Carlos Red-on-brown and Safford Variety Middle Gila Buff Ware. In addition, measures of richness, which calculate the diversity in a population, demonstrated that the decorated ceramic assemblages of sites postdating A.D. 1275 were substantially more diverse than the assemblages of sites occupied before A.D. 1275. Analyses of the spatial distribution of decorated ceramics also illustrated that the distribution of San Carlos Red-on-brown, a type local to the Safford and Aravaipa valleys, was often spatially segregated from the distribution of Maverick Mountain Series types, which are associated with puebloan immigrants (Neuzil 2005b).

Petrographic analyses allowed an examination of the provenance of decorated ceramics, to evaluate whether non-local decorated ceramics were traded into the area, and thus not evidence of migration, but rather exchange, or were produced locally by migrants who moved into the area and brought knowledge of ceramic technology and design style from their homelands with them. These analyses revealed that many ceramic wares, including those associated with migrant populations such as Maverick Mountain Series and Roosevelt Red Ware, were manufactured in the Safford and Aravaipa valleys with local raw materials. Thus, migrants with the technological knowledge to create these non-local ceramics were likely living in the Safford and Aravaipa valleys after A.D. 1275 (Neuzil 2005b).

In total, the ceramic evidence from many sites in the Safford and Aravaipa valleys points towards distinct households with unique enculturative backgrounds and the presence of local production of ceramics typically associated with both immigrants and indigenous groups. Coupled with evidence of migrant populations living at the Goat Hill site, this ceramic evidence demonstrates that migrants from the Kayenta and Tusayan areas of northeastern Arizona were present in the Safford and Aravaipa valleys after A.D. 1275 and lived side-by-side with indigenous groups at many late Classic period settlements in this area, as evidence of both groups were found at almost all sites occupied during this time.

Several lines of evidence from domestic architecture also point towards coresidence of migrant and indigenous groups after about A.D. 1275. First and most compelling is the use of cobble reinforced adobe (see Figure Nine), a local construction technique, to build room blocks, a puebloan architectural layout. All sites primarily occupied after A.D. 1275 were constructed in this manner. Room blocks are generally more common in areas to the north, where they are regularly constructed of horizontally coursed masonry. Their appearance in the Safford and Aravaipa areas after approximately A.D. 1275 signals a profound shift in the grammar of site layout and arrangement of space that is likely associated with the arrival of migrant populations from the Kayenta and Tusayan areas. This cannot be attributed to raw material availability, as two sites in the Safford Valley, the Goat Hill site and the Marijilda site (AZ CC:5:6[ASM]), were constructed with masonry. Combined with other data, these patterns suggest that both migrants and indigenous populations influenced settlement layout and construction, and indicates that these populations coresided in the same settlement shortly after the arrival of migrants in the area (Neuzil 2005b).

This evidence is paralleled by a general shift in the Room Contiguity Index (RCI), which measures the number of rooms at a settlement that share walls by dividing the number of walls by the number of rooms present at a site (Clark 2001: 46-56). This index was created to differentiate between sites that exhibit an open compound layout with a dispersed room arrangement, and those arranged as contiguous room blocks. RCI values can range from 2.0 (a square room block of infinite size) to 4.0 (a site with no contiguous rooms), and sites dominated by non-contiguous rooms have higher RCIs, while sites dominated by rooms that share walls have lower RCIs. Clark's (2001) regionwide assessment of sites in the Southwest demonstrated there were two distinct architectural traditions during the thirteenth century, the Ancestral Puebloan masonry room blocks of the northern Southwest, and the Hohokam adobe compounds of southern Arizona.

Figure 9: Cobble reinforced adobe.

The pattern in RCI values for the Safford and Aravaipa valleys was mixed. While most early Classic period settlements in the Safford and Aravaipa valleys were arranged into compounds with single rooms attached to an outer compound wall, yielding high RCI values, after A.D. 1275 (see Figure Ten), when room blocks became the dominant architectural form, these values became lower, indicating increasing use of architectural layouts with rooms sharing walls, as is the case in a room block arrangement. Thus, RCI values from Safford and Aravaipa area sites suggest that both puebloan migrants and indigenous groups had a substantial influence over site layout and construction materials, indicating both groups likely lived together at these settlements (Neuzil 2005b).

Measures of space syntax, which provide a method of understanding social behavior by quantifying differences in the use and structure of space, were also applied to architectural data from Safford and Aravaipa valley sites. The underpinning principle of space syntax analyses is that spatial integration is a proxy for social integration (Ferguson 1996, Hillier and Hanson 1984; Potter 1998), and can thus be used to understand how people interact through the way they structure the spaces in which they live. Analyses of the open space at Safford and Aravaipa valley sites showed that settlements postdating A.D. 1275 were not as well spatially and socially integrated as those dating before A.D. 1275, but were more spatially and socially integrated than migrant enclave sites (the Goat Hill and Marijilda sites, and AZ CC:2:23[BLM]) dating to this same interval (Neuzil 2005b). This suggests migrant and indigenous groups who chose to live together purposefully created settlement structures that encouraged integration and social cohesion, but were somewhat limited by the existing social space between them.

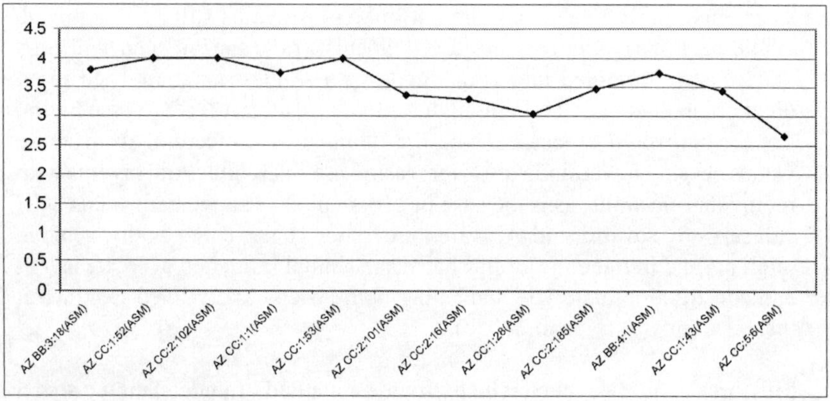


Figure 10: Average RCI for select sites. All sites to the right of AZ CC:2:101(ASM) postdate A.D. 1275.

As this discussion demonstrates, evidence from multiple sites in the Safford and Aravaipa valleys suggests that Puebloan migrants arrived during the late thirteenth and fourteenth centuries both as suprahousehold level groups that established their own settlements apart from the local population, as at the Goat Hill site, and as household level groups that settled alongside the local population at many other late Classic period sites. This pattern indicates that the size of the migrant group may have, in part, influenced whether migrants chose to establish a settlement removed from the local population or not. While it is apparent that the initial wave of Puebloan migrants arrived and settled as suprahousehold level groups, it is still unclear whether subsequent migrants arrived in the Safford and Aravaipa valleys as larger suprahousehold level groups that then fissioned into constituent households, or arrived as individual households.

## The Effect of Migration on Identity

The impacts of migration on identity in the Safford and Aravaipa valleys appear to have been largely the result of the level of integration between migrant and indigenous groups. The first groups of migrants to the Safford and Aravaipa areas who arrived during the late thirteenth century appear to have initially remained physically and socially separated from the local population. Excavations have demonstrated that the people living at the isolated and defensible Goat Hill site were almost exclusively migrants, based on the ceramics and architectural elements found at the site (Woodson 1995, 1999). Other potential migrant enclaves, such as AZ CC:2:23(BLM), the Smith Tank site, and the Marijilda site, exhibit many of the same characteristics of physical isolation in their locations, and social isolation evidenced in the material culture, specifically ceramics and architecture, found at each. Immediately after arriving at their destination, migrants undoubtedly still identified themselves as being socially distinct from the local population in the area, and vice versa.

However, after the initial arrival and settlement of these migrant enclaves, this separation appears to have slowly become less pronounced. On the whole, evidence from early fourteenth century sites suggests that migrant and local groups later came together to form additional settlements that included members of both populations. Both groups contributed to these newly founded settlements, as is demonstrated in the architecture at these sites, which represent a combination of local and migrant construction techniques. However, it appears that at the same time each group maintained some differences, evidenced by the continuation of distinct decorated ceramic traditions found side-by-side at many of the sites from this time. Thus, the evidence suggests migrant and indigenous groups renegotiated their identity and redefined their social space in order to facilitate coresidence, but likely would not have characterized themselves as a single cohesive social group at this time.

During the later fourteenth century, the pattern in architectural layout and construction techniques continued the developments seen previously. During this time, sites in the Safford and Aravaipa valleys became significantly larger as fewer settlements were constructed, and those that were built and occupied during this time became increasingly nucleated. The overall layout of sites also changed somewhat, as the larger sites were constructed as contiguous and compact roomblocks, often surrounding plazas. These developments, combined with an increase in the number of people living at each site, undoubtedly required the development of new integrative institutions that would ensure the inhabitants could live together peacefully and functionally.

The totality of evidence from sites occupied after A.D. 1325 suggests that while Puebloan migrants became increasingly integrated into the local population during this time, each group still maintained social networks and exchange patterns established before migrants arrived in the Safford and

Aravaipa valleys. Sites continued to demonstrate a mixture of Puebloan room block layout with local cobble reinforced adobe construction techniques. However, overall social integration had most certainly increased since the arrival of migrants in the area in the late thirteenth century, seen in the decrease in diversity among decorated ceramic assemblages, dominated by Roosevelt Red Ware, at later fourteenth century sites. Previous research has suggested that Roosevelt Red Ware vessels were initially produced by migrants from the Kayenta and Tusayan areas in many areas of the Southwest in order to sustain themselves via exchange of ceramic vessels given their limited access to resources and good agricultural land as latecomers to the areas in which they settled (Crown 1994:213; Lyons 2003:66). Production of this ware likely spread to other social groups thereafter, given that Roosevelt Red Ware vessels are found at sites with no strong indicators for Kayenta and Tusayan migrants (Crown 1994:209; Lyons 2003:66). Therefore, their dominance at fourteenth century sites suggests almost all sites dating to this time were either inhabited at least in part by Kayenta and Tusayan migrants, or they were involved in regular exchange or interaction with Kayenta and Tusayan migrants. Furthermore, given the increase in site size and the decrease in the number of sites as settlements became increasingly nucleated in the Safford and Aravaipa valleys, it is most likely that Kayenta and Tusayan migrants and indigenous groups were living together at the same settlements during this time. However, the number of migrants and indigenous groups may not have been equal at each settlement, establishing a power differential between these two groups. In addition, since evidence from late thirteenth century and early fourteenth century sites suggests migrant and local groups began living together shortly after the arrival of the migrants, it is likely that some of the social boundaries between these groups steadily became increasingly permeable through intermarriage and daily social interaction.

The totality of evidence from sites occupied after A.D. 1325 suggests that while Kayenta and Tusayan migrants became increasingly integrated into the local population during this time, each group still maintained social networks and exchange patterns established before migrants arrived in the Safford and Aravaipa valleys. Sites became larger and more nucleated, but continued to demonstrate a mixture of puebloan room block layout with local cobble reinforced adobe construction techniques. Measures of space syntax remained variable during this time, suggesting that social integration across the region was still not uniform. This is further reflected in decorated ceramic assemblages, which remain variable through the fourteenth century, demonstrating a maintenance of previous social networks for the production and exchange of ceramics at this time. Despite this fact, overall social integration had most certainly increased since the arrival of migrants in the area in the late thirteenth century, seen in the decrease in diversity among decorated ceramic assemblages at later fourteenth century sites. In this unique situation where the structure underlying the expression of the social identities of both migrant and indigenous groups was challenged, individuals in each of these groups chose to alter this structure of identity expression to fit in, socially, with their new neighbors. Such alterations of practice are reflected clearly in the archaeological record (Gardner 2002; Meskell 2002).

## Social Consequences of Migration in the Safford and Aravaipa Valleys

Based on the evidence summarized here, it appears Puebloan migrants arrived in the Safford and Aravaipa valleys in both household and suprahousehold level groups. When they arrived, the local population was organized at a similar social scale, based on the size and number of sites dating to the earlier thirteenth century (Bylas phase; see Figure Two). Immediately after the arrival of the first suprahousehold level migrant groups that settled at the Goat Hill site, and probably also at AZ CC:2:23(BLM), the Smith Tank site, and the Marijilda site, migrants and indigenous populations appear to have maintained their separate social identities and lived spatially segregated from each other. Shortly thereafter, household level migrant groups began settling in the valleys as well, and appear to have chosen to live alongside local populations who were organized into larger or comparably sized social groups. At this time, both migrant and indigenous groups renegotiated their identities to create a new identity that incorporated aspects of each pre-migration identity. Thus, the manner in which each of these groups expressed their identity shifted to reflect a newly defined structure delineated by the social realities of the post-migration environment.

In sum, the Safford and Aravaipa valleys have provided an excellent case study to assess the social consequences of migration. While little research in the Southwest has addressed such consequences directly, they are clearly crucial to understanding the social environment of the last centuries before European contact. This research provides a framework upon which future research on migration can be based.

## Notes

[1] Archaeological sites in Arizona are referenced by site numbers and often by names. The site numbering system in widest use in Arizona is that of the Arizona Site Survey at the Arizona State Museum. The number includes a state abbreviation prefix (some sites recorded by the Arizona Site Survey are located in California or in the Mexican states of Sonora or Chihuahua), a series of numbers that identify the site uniquely within subdivisions of 1 degree longitudinal blocks, and a suffix that identifies the organization that issued the number. The sites discussed in this volume are recognized by the Arizona State Museum (ASM) or the U.S. Department of the Interior, Bureau of Land Management (BLM). Due to the cumbersome nature of this scheme, site names are used when they are available, but not all the sites in the Arizona Site Survey have been named.

# CHAPTER TWO

# MULTI-CULTURAL COMMUNITIES IN THE SAFFORD BASIN

DAVID E. PURCELL AND JEFFERY J. CLARK

The existence of prehistoric immigrant communities in the Safford Basin achieved high visibility in the archaeological literature only during the past decade, largely due to the work of the University of Texas at Austin (Woodson 1995, 1999). Woodson's investigation of the Goat Hill Site (AZ CC:1:28[ASM]) revealed a small enclave of Kayenta/Tusayan migrants from northeastern Arizona. The Goat Hill site appears to date to the late thirteenth century A.D. (all dates are A.D. unless otherwise noted) when communities across the Greater Southwest began to coalesce.

Two projects that were conducted in the Safford Basin during 2000-2001 revealed the existence of examples of multi-cultural communities in which local people and immigrants developed closer relationships with one another than that implied by the remote and defensive location of the Goat Hill site. In these recent cases, "Hohokam-influenced" groups appear to have interacted in distinctly different ways with ancestral Puebloan groups from the Mogollon highlands and northeastern Arizona. Both projects, funded by the Arizona Department of Transportation, exposed tantalizing slices of culturally diverse villages.

From September 2001 through January 2002, Archaeological Research Services, Inc. (ARS) conducted excavations within portions of two extensive village sites south of Safford along U.S. Highway 191 (Purcell 2004a). The two sites—The Artesia Site (AZ CC:6:40[ASM]) and Roadrunner Village (AZ CC:6:43[ASM])—are on the eastern bajada of the Pinaleño Mountains below Mount Graham. Both sites witnessed long-term, intensive occupations, but the Artesia Site contained only four intact features within the project area. In contrast, many intact features were found at Roadrunner Village, with superimposed and remodeled residences testifying to the permanent occupation at that site.

In Spring 2001, Desert Archaeology, Inc. (DAI) conducted excavations along U.S. Highway 70 between Safford and Thatcher (Clark 2004). Eight prehistoric sites were investigated, including three prehistoric settlements, one prehistoric field, one historic canal segment, and two prehistoric canal segments (see Chapter Three, this volume). At the Daley site (AZ CC:2:235[ASM]), a short-term migrant encampment was identified at the margin of a local village previously investigated by Eastern Arizona College ([EAC] 1981)

As is often the case with highway projects, only bits and pieces of sites from each project were available for investigation. However, the evidence from Roadrunner Village and the Daley Site, when viewed with examples from other excavated sites in the Safford Basin, indicate complex and multilateral relationships had developed among the various cultural or ethnic groups residing in the region by the late thirteenth century. This observation is potentially significant in considering the subsequent status of immigrant populations during the fourteenth century, a time of dramatic change and rapid settlement aggregation in the Safford Basin (Clark et al. 2004; Neuzil 2005b)

Previously, Purcell (2004a) proposed that four distinctive relationships between local villagers and recent immigrants can be observed in the ancient Safford Basin during the Ceramic period of prehistory:

1. culturally segregated communities or enclaves (e.g, the Goat Hill Site [Woodson 1995, 1999]);
2. co-resident settlements (Whittlesey and Heckman 2000:10) that include more than one culture group (e.g, the Spear Ranch [AZ CC:1:11(ASM)]/Krider Kiva [AZ CC:1:43(ASM)] complex [Neuzil 2005b];
3. settlements of people with a persistent local tradition that was influenced by other sociocultural groups without substantial migration (e.g., the San Simon Mogollon [Sayles 1945]), and
4. ethnogenesis, the emergence of new tradition that represents a blending or mixing of previously distinctive cultures (e. g., the Western Pueblo concept of Johnson and Wasley [1966] or the

Salado, to some researchers [Brown 1973; Dean 2000]) .

Two different examples of co-residence are examined here. Roadrunner Village exhibits both Mogollon and Hohokam attributes, suggesting possible Hohokam integration into a local (?) "San Simon" Mogollon community, perhaps through marriage. At the Daley site, Kayenta/Tusayan immigrants camped on the fringes of a large village that may have included groups from both Hohokam and highland Mogollon cultural backgrounds. These immigrants either moved on or where ultimately accepted into the village.

## Hohokam-Mogollon Coresidence at Roadrunner Village

Roadrunner Village is located south of Safford near natural artesian springs, on the lower bajada slopes of the Pinaleño Mountains. The site is large, measuring at least 640 m north-south along U.S. 191, and at least 412 m east-west. Most of the site is located on private land that borders the highway. The site has been known informally since the 1970s, when Betty Graham Lee of Eastern Arizona College reported its existence and named it the Roadrunner site after a now defunct nearby restaurant. ARS encountered the site during a survey of the U.S. 191 right-of-way in advance of this project, and was able to trace it away from the highway by observing the distribution of artifacts in the roadside ditches of residential streets in the area (Wright 2001). Some of the local residents, many of whom are amateur archaeologists, informed ARS of the existence of adobe compounds within the site boundaries, but beyond the narrow project area (25 feet [8 m] on each side of the existing road).

### Results of the U.S. 191 Investigations

ARS excavated the intact portions of 48 features, including 17 habitations that exhibited variable architectural forms (Figure Eleven). Circular, D-shaped, rectangular, square and irregularly shaped pithouses were excavated, as well as a large masonry structure that could have been a surface room or even part of a compound. The upper portion of the walls had been previously removed by highway construction. Although its true nature could not be determined, this room yielded nearly one third of all artifacts recovered from this site, including 10,000 sherds. Most of the houses did not have culturally distinctive architectural attributes, but were consistent in shape and construction with those included within the local

"San Simon" Mogollon tradition (Sayles 1945; Gilman 1997), although several houses included non-local architectural details such as a raised floor, house-in-pit construction, and adobe-stepped entries more commonly associated with Hohokam dwellings (Haury 1976; Motsinger 1994; Seymour 1994).

Occupation must have been year-round, as indicated by the substantial construction of the habitations and presence of numerous large storage pits, although few hearths were identified (due to the narrow project area, sampling strategy, and disturbance). Pollen and macrobotanical remains also indicate use of the site from at least the spring through fall months. Two features at Roadrunner Village were specialized storage features, perhaps granaries or animal pens. These were constructed from what originally had been large subfloor storage pits within a pithouse, and rebuilt as above ground jacal silos or pens surrounded by post enclosures.

The earliest feature at Roadrunner Village dates to circa A.D. 600-650 in the Dos Cabezas phase, using the preliminary chronological scheme proposed by Clark et al. (2004) and Crary (1997) (see Figure Two, this volume). Roadrunner Village appears to have been occupied during two long intervals from A.D. 600-900 or 950, and from A.D. 1000 to after 1200, although many features are imprecisely dated and could fall within the apparent hiatus (Purcell 2006a). The second occupation contains a greater number of features, with evidence of continuous occupation through that interval, which spans the Two Dog and Eden phases, ending during the Bylas phase. The possible surface room or compound (Feature 15), is the only intact feature found by ARS with evidence for use after A.D. 1150 to 1200. Roadrunner Village appears to have once contained a substantial Classic period occupation, but evidence for this was removed by modern construction within the right-of-way or is on private property elsewhere in the site. Feature 15 could have been part of a compound, such as those that appear to be typical of the Bylas phase Owens-Colvin (Rule 1993) and Daley sites (EAC 1981).

## Material Evidence of Coresidence

During excavations at Roadrunner Village, pithouse superpositioning and variable architectural forms and attributes suggested the possibility that the site had been occupied by peoples from different cultural backgrounds. This possibility was further examined during the artifact analysis (Purcell 2004b;

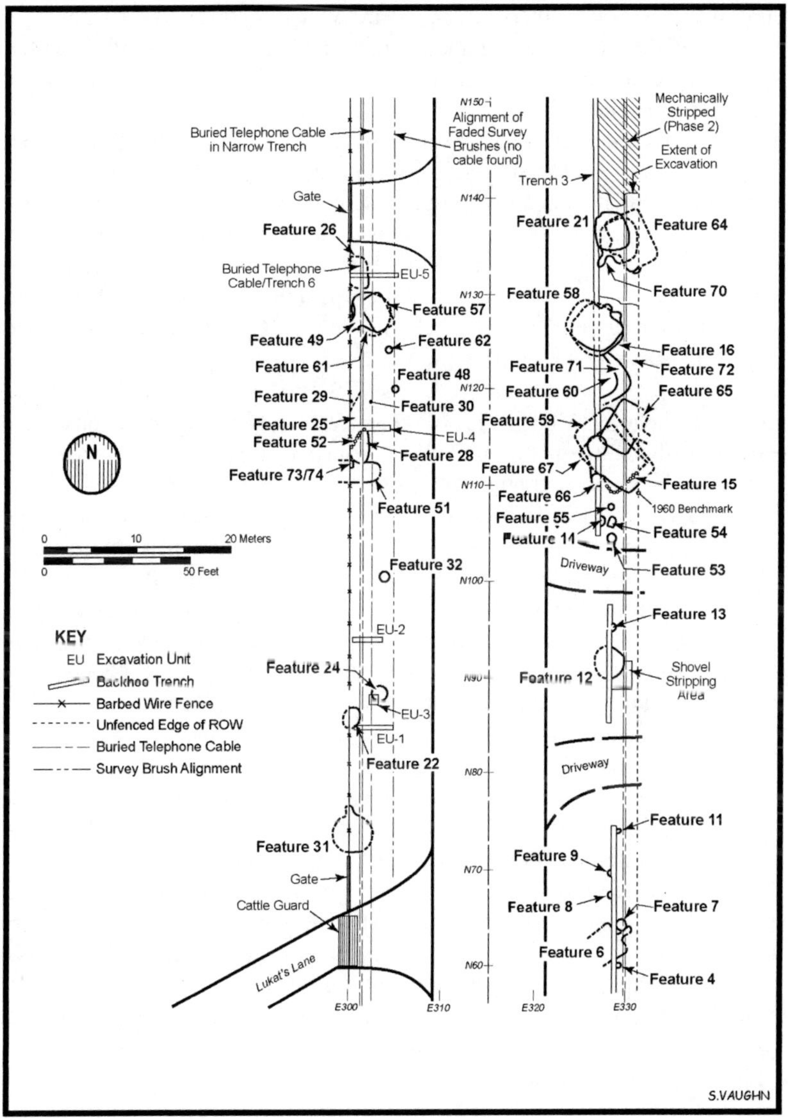


Figure 11: Excavated features in Locus A at Roadrunner Village, illustrating the variability of house forms.

Purcell and Coriell 2004a, 2004b; Smith et al. 2004). Assessments of cultural affiliation were conducted for each feature to avoid potential problems with temporally mixed assemblages (Purcell 2006a:75-78). Certain architectural features, ground stone artifacts, ceramics, jewelry or items of personal adornment, and human remains indicate that Roadrunner Village was a typical "San Simon" Mogollon village that had increasingly close contact with Hohokam groups to the west through time from A.D. 600 to at least A.D. 1150/1200 (Purcell 2006a). Based on the types of Hohokam ground stone and personal adornment, the presence of cremated human remains, and evidence for on site pottery production using the Hohokam paddle-and-anvil method, Purcell (2004a, 2006a) hypothesized that Hohokam women had married into the village.

ARS was unable to determine if genetically different populations were present at the settlement. No formal burials were identified or excavated and all of the human remains were found in secondary deposits with other disturbed cultural fill. In addition, invasive methods of analysis were not permitted in the project burial agreement. The distinction is important to addressing the question of whether the archaeological "Hohokam" represent a biologically distinct group who migrated across southern Arizona, or were merely a sociocultural construct that included members of multiple groups who adopted Hohokam ideology and lifeways. The evidence of co-resident Hohokam at Roadrunner Village is discussed in detail below, following Purcell 2004 and 2006a.

*Architectural Evidence*. All of the domestic architecture is typical of "San Simon" Mogollon forms (Sayles 1945), irregular and less formal than most Hohokam residential architecture. The house (Feature 6) with the most distinctively Mogollon architecture is a deep square or rectangular house pit with a long, narrow entry ramp, and is very similar to Galiuro Phase houses excavated by Sayles at San Simon Village. All but two of the houses (16 and 21) are true pithouses, where the house pit forms the inner wall of the dwelling. A few Hohokam architectural details were present: a formal adobe-lined and coped hearth, and short entryway with adobe step in Feature 64; a raised floor in Feature 65; and, house-in-pit construction (Features 16 and 21). Features 64 and 65, however, were true pithouses, and Features 16 and 21 were otherwise irregular in plan and informal in construction (Figure Eleven). Other Hohokam architectural details such as wall grooves or Hohokam floor plans (Haury 1976; Motsinger 1994) were not observed at Roadrunner Village.

*Ceramic Evidence.* The ceramic assemblage from Roadrunner Village included large quantities of both Hohokam and Mogollon pottery, indicating close ties between these groups (Smith et al. 2004). Most of the utility wares at each site was made in the Mogollon coil-and-scrape technological tradition, and is one of the most convincing lines of evidence that the majority of the population was of "San Simon" Mogollon affiliation. Mogollon ceramics included all corrugated types, Mogollon Utilitarian Ware, "Other" Red-on-Brown pottery (mostly San Simon and Mogollon highland types), and Mimbres Black-on-white and Mimbres Polychrome. Distinctively Mogollon pottery represents 9.98 percent of the ceramic assemblage at Roadrunner Village, with a ratio of 2:1 Mogollon to Hohokam pottery (Table Three). Securely dated features exhibited a range from 15.5:1 (Feature 67) to 1.3:1 (Feature 24). Relatively large quantities of Mimbres pottery (although <2.0 % of the assemblage) was probably obtained through trade with Mimbres settlements that are known from the eastern side of the San Simon valley and Gila Mountains.

Hohokam presence was measured by the percentage of all buff wares in the total assemblage (combining those produced in the Middle Gila and Safford areas), and mica schist-tempered plainware and redware sherds. Smith et al. (2004) associate the use of this temper with production in the Middle Gila valley in the Hohokam heartland. Temper analysis indicates that buff ware vessels recovered from Roadrunner Village were made both in the Middle Gila region, and locally in the Safford Basin. Regardless of the origins of the buffware found at Roadrunner Village, Hohokam pottery was readily available to the residents of this village, representing up to 8.92 percent of individual feature ceramic assemblages. Some of this pottery may have been made at Roadrunner Village, as indicated by the presence of numerous heavily worn polishing stones and the possible pottery anvil (Figure Twelve); Mogollon pottery is built by coiling and scraping pottery, and would not require the use of a pottery anvil (Crown 1991). Anvil marks are common on plainware jars from the site, a type of pottery considered less likely to be exchanged among different groups (Smith et al. 2004). San Carlos Red-on-brown was excluded from this comparison because it exhibits a mix of Hohokam and Mogollon pottery attributes, and post-dates the primary period of occupation at Roadrunner Village (Smith et al. 2004).

**Table 3: Ratios of Hohokam and Mogollon Ceramics Recovered from Roadrunner Village Features.**

| FEATURE | | | CERAMIC ASSEMBLAGE | | | |
|---|---|---|---|---|---|---|
| No. | Type | Date (A.D.)[a] | Total | Hohokam %[b] | Mogollon %[c] | Ratio H:M[d] |
| 65 | Pitstructure | 600-650 | 455 | 1.31 | 5.93 | 4.5 |
| 61 | Pitstructure | 600-750 | 99 | 2.02 | 14.14 | 7.0 |
| 31 | Pitstructure | 650-900 | 287 | 2.09 | 3.14 | 1.5 |
| 67 | Storage pit | 700-900 | 139 | 1.43 | 22.30 | 15.5 |
| 58 | Pitstructure | 1000-1050 | 565 | 2.65 | 8.31 | 3.1 |
| 70 | Pitstructure | 1000-1050 | 49 | 6.12 | 4.08 | 5:1 |
| 24 | Storage pit | 1000-1100 | 402 | 5.72 | 7.71 | 1.3 |
| 12 | Pitstructure | 1000-1100 | 587 | 3.57 | 7.84 | 2.2 |
| 59 | Pitstructure | 1000-1100 | 750 | 1.70 | 8.53 | 4.9 |
| 57 | Pitstructure | 1040-1100 | 369 | 4.60 | 6.78 | **1.4**[e] |
| 21 | Pitstructure | 1050-1100 | 112 | **8.92**[e] | 19.64 | 2.2 |
| 16 | Pitstructure | 1050-1130 | 1658 | 3.19 | 7.17 | 2.4 |
| 49 | Pitstructure | 1100-1160 | 2737 | 6.97 | 12.31 | 1.7 |
| 64 | Pitstructure | 1100-1160 | 259 | 3.86 | 11.58 | 3.0 |
| 15 | Surface room | 1160-1200+ | 9913 | 5.80 | 12.99 | 2.2 |

[a] See Purcell 2006 for derivation of dates
[b] All buffware sherds; and mica schist-tempered plainware and redware.
[c] All corrugated, Mogollon Utilitarian Ware, Mimbres Black-on-white and Polychrome, and "other" Red-on-brown (Smith et al. 2004).
[d] Hohokam:Mogollon ratio, 1:×
[e] Highest percentage of Hohokam ceramics and lowest ratio to Mogollon ceramics

*Bioarchaeological Evidence*. ARS collected 392 fragments of human bone, representing 385 individual skeletal elements, from Roadrunner Village (Purcell and Coriell 2004b). Twenty-two prehistoric features contributed 94.0 percent (*N*=362) of the human bone recovered from this site, with only 4.9 percent (*N*=19) obtained from the surface and 1.0 percent (*N*=4) obtained from an excavation unit (Purcell 2004: Table 11.5). Burned or cremated bone was 84.8 percent (*N*=328) of the site assemblage, suggesting that cremation was the preferred means for disposing of the dead at this site as well. Cremation of the dead is strongly associated with the pre-Classic Hohokam (Haury 1976) and generally not with the Mogollon (Wheat 1955).

The majority of the bone came from four features, including two pithouses (Features 6 and 16), a masonry structure (Feature 15), and an oval pit (Feature 11). All of the bone from Features 6 and 16 had been burned, probably in cremation rites. Based on the composition and provenience of ground stone artifacts, Features 6 and 16 seem to have contained at least partial floor assemblages, with trash filling the house pits above the floor. The human bone from these features, however, was distributed throughout the fill of each house, suggesting that they were deposited within the houses with other trash. It appeared that, rather than human bone being intentionally discarded with domestic refuse, later house construction and remodeling resulted in the disturbance of nearby trash middens and burials, with the mixed materials being dumped into open house pits and other abandoned features. Most of the human bone found in the fill of Feature 15 was burned (96.0 %). The composition of the artifact assemblage from this feature suggests that the large feature had been used for trash disposal by people associated with a Late Formative/Classic period occupation at the site from which no intact features were excavated, and the human bone was part of this deposit. Feature 11, a small oval pit, is the only feature of its shape excavated at Roadrunner Village. During its investigation, ARS excavators hypothesized that it was a burial pit due to the unusual shape, the presence of large quantities of bone (*N*=52; 14.4 %), and apparent burned condition of the pit. Unfortunately, it had been heavily disturbed by rodents and/or vandals. Most of the human bone (98.1 %), had not been burned, however, suggesting that the pit could have held an inhumation, not a cremation.

*Ground Stone Evidence.* The excavated features at Roadrunner Village contributed a large assemblage with many specialized Hohokam forms, including a large number of trough metates, formal slab metates, one-hand manos, formal pestles, a decorated palette fragment, stone balls, a scoria censer fragment, ¾-grooved axes, a "chunkee" (Haury 1976) or doughnut stone, a hoe, and a turquoise mosaic tessera (Table Four, Figure Twelve). ARS also recovered a fragment of a possible cylindrical mortar and a possible pottery anvil, both types associated with the Hohokam and not the Mogollon. The distribution of Hohokam ground stone is mixed with Mogollon forms in all but one dated feature (27), which contained a single one-hand mano (Purcell 2004b). Furthermore, the three dated features with distinctively Hohokam ground stone artifacts (Features 31, 16, and 15) also yielded substantial numbers of Mogollon ground stone artifacts. For example, Feature 15 had 13 Hohokam types (including the censer and torus fragments) and 45 Mogollon types. Only two features (49 and 15) contained complementary Hohokam hand and nether stones, consisting of slab metates and one-handed manos in both cases. The admixture of Hohokam and Mogollon artifacts is one of the attributes of the assemblage that suggests co-residence of individuals at the household level, rather than the presence of distinct Hohokam households or larger social units. Although the specialized Hohokam artifacts could be attributed to trade, the social context in which these forms were used and consumed suggest that they were locally produced with the possible exceptions of decorated palettes and censers probably used in cremation rituals (Haury 1976:288; Wilcox 1991). In particular, the pottery anvil would be of little use to Mogollon potters.

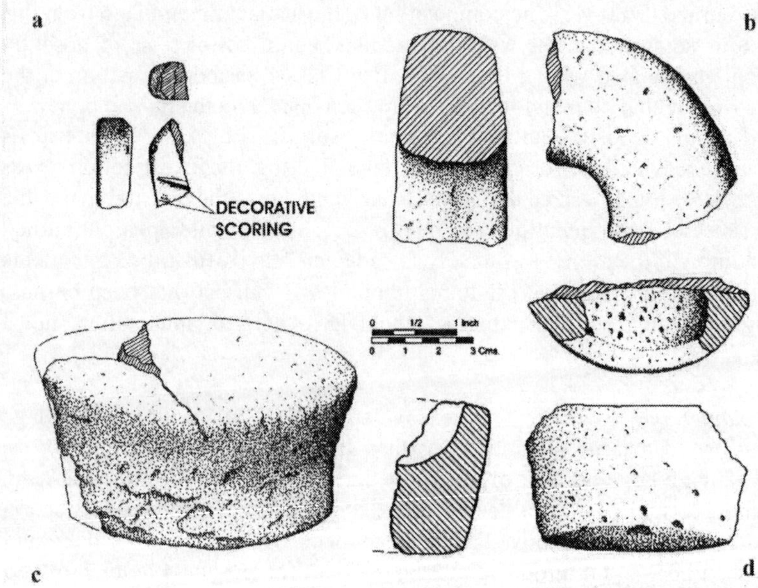


Figure 12: Hohokam ground stone forms recovered from Roadrunner Village. (a) Three views of a fragment of a small, formal pallet. (b) Two views of a "chunkee" or doughnut stone. (c) Pottery anvil. (d) Three views of a stone censer.

## Table 4: Mogollon and Hohokam Ground Stone Forms at Roadrunner Village.

| TYPE | | MOGOLLON PERIOD San Simon Branch (Wheat 1955) | | | | HOHOKAM PERIOD At Snaketown (After Haury 1976) | | ROAD-RUNNER VILLAGE |
|---|---|---|---|---|---|---|---|---|
| | | 1 | 2 | 3 | 4 | Preclassic | Classic | |
| *Netherstones* | | | | | | | | |
| Metates | Basin | P | P | P | P | 2 | - | 7 |
| | Basin/trough | P | ? | P | P | - | - | - |
| | Full Trough | - | - | - | P | 103 | 3 | 10 |
| Other | Grinding slab | C | P | P | P | - | - | 8 |
| | Grinding basin | P | P | P | P | 3 | 1 | 7 |
| | Palette | C | - | P | P | 49 | - | 2 |
| | Non-grooved abrader | P | P | P | P | P | P | 3 |
| | Grooved abrader | - | - | - | - | P | P | 2 |
| Mortars | Small flat/disk | - | - | - | - | - | - | 1 |
| | Gyratory crusher | - | - | - | - | 1 | - | - |
| | Slab | - | - | - | - | - | 1 | - |
| | Cylindrical | - | - | - | - | 1 | - | 1 |
| *Handstones* | | | | | | | | |
| Manos | One-hand | - | - | - | - | 26 | 8 | 16 |
| | Two-hand | - | - | - | - | 160 | 13 | 2 |
| | One-or two-hand | P | C | A | A | - | - | 25 |
| | Handstone | A | C | C | P | 34 | 11 | 64 |
| Pestles | Faceted | C | P | P | P | - | - | - |
| | Informal | P | - | P | P | 2 | 3 | 4 |
| | Formal | - | - | - | - | 3 | 11 | 3 |
| *Other* | | | | | | | | |
| Axes | ¾-grooved (ridged) | - | - | - | - | 25 | - | 2 |
| | ¾-grooved (unridged) | - | - | R | - | 21 | 6 | 1 |
| Censers | Plain | - | - | - | - | 11 | - | 1 |
| | Decorated | - | - | - | - | 114 | - | - |
| Stone rings | Plain doughnut/torus | - | - | - | R | 4 | - | - |
| | Pulley doughnut/torus | - | - | - | - | 7 | - | - |
| | Chunkee/bi-concave | - | - | - | - | P | P | 1 |
| | Blanks | - | - | - | - | 11 | 2 | - |
| | Finger ring | - | - | - | - | P | P | 2 |
| Stone balls | | - | - | - | - | 3 | 2 | 2 |
| Pottery anvil | | - | - | - | - | - | - | 1 |
| Hoe | | - | - | - | - | 4 | 10 | 1 |
| Beads (stone & shell) | Disk | - | - | - | - | 132 | 6 | 22 |
| | Bi-lobed/figure-8 | - | - | - | - | 11 | - | 1 |
| | Barrel | - | - | - | - | 3 | - | 3 |
| | Tube | - | - | - | - | | | 1 |
| | Whole shell | - | - | - | - | 59 | 5 | 1 |
| Pendants | Geometric | P | P | - | P | 16 | - | 7 |
| | Effigy | - | - | - | - | 1 | 0 | - |
| Tessera | | - | - | - | - | 32 | 2 | 1 |

*Notes:* A=abundant; C=common; P=present; R=rare; ?=probable but not documented

*Jewelry*. Shell and stone jewelry were relatively common at Roadrunner Village, with most shell artifacts being *Glycymeris* sp. bracelet fragments. Most of the forms, such as *Glycymeris* bracelets, are common at both Hohokam and Mogollon sites (Table Four). However, disk, bi-lobed, barrel, tube, and whole shell beads, and effigy pendants were commonly recovered at Snaketown in the Middle Gila Valley (Haury 1976), but not at San Simon sites (Wheat 1955:147).

*Summary*. Various lines of evidence regarding the cultural affiliation of the inhabitants of Roadrunner Village—including most of the ceramics, ground stone, and architectural features—lead to the conclusion that this is a local "San Simon" Mogollon settlement. However, a relatively large number of Hohokam artifacts and architectural traits suggest that some Hohokam may have been living in Roadrunner Village, where they may have married into the community. Purcell (2004a, 2006a) has hypothesized that at least some of the buffware vessels of local manufacture were made on-site and used by Hohokam spouses of Pinaleño bajada villagers. The artifacts and architecture of this village indicate persistent contacts with the Middle Gila Valley beginning (possibly) as early as A.D. 600-650. By A.D. 1000 Roadrunner Village and the Owens-Colvin site (AZ CC:1:19[ASM]) display very similar patterns in the occurrence of Hohokam artifacts, most of which, in the words of Pam Rule (1993:51), "reflect ritual or technological conventions of a type unlikely to diffuse unaltered across traditional cultural lines," suggesting that Hohokam intermarriage into Mogollon villages (if that is the mechanism) was widespread in this part of the Pueblo Viejo District.

## A Kayenta/Tusayan Camp on the Edge of the Daley Site

On the eastern edge of Thatcher, DAI investigated the south margin of the Daley site, a large Bylas phase (see Figure Two) village, in a linear swath 8 m on either side US Highway 70 (Figure Thirteen). A small portion of the central site area was sampled by Eastern Arizona College in the early 1980s as part of a salvage operation associated with laser terracing of agricultural fields (EAC 1981). This fieldwork included the excavation of portions of at least one masonry compound, occupied by local groups perhaps of mixed Mogollon-Hohokam archaeological affiliation similar to that identified at Roadrunner Village discussed above. DAI's narrow exposure of the Daley site revealed a portion of an ephemeral late thirteenth century pitstructure component (Figure Fourteen). Evidence for other activities that occurred on the settlement margins such as trash disposal, and perhaps ceramic manufacture, was also identified in a deposit filling a small wash that was open during the late thirteenth century. These activities were probably undertaken by the permanent residents of nearby compounds such as the example tested by Eastern Arizona College.

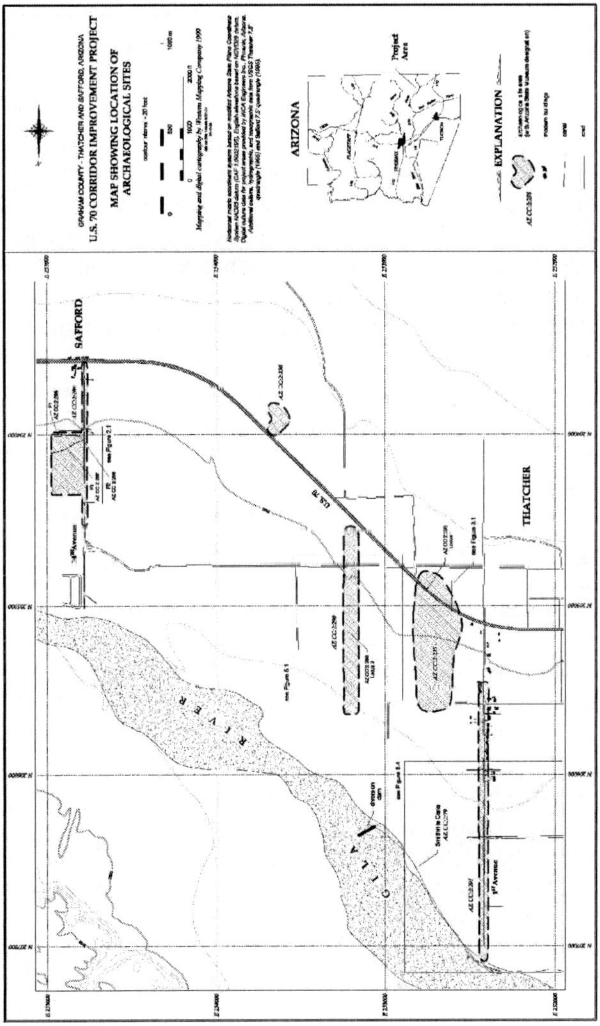

Figure 13: Locations of sites investigated by Desert Archaeology, Inc. during the U.S. 70 Project in Thatcher, Arizona.

In the small portion of the Daley site available for investigation by DAI, remarkably different, yet contemporaneous, decorated ceramic assemblages were recovered from the pit structure component and wash, less than 100 m from each other (Clark 2004:Table 3.2). All 206 Maverick Mountain series sherds recovered from the site were from the fill and floor of two superimposed and insubstantial pit structures (Features 1 and 9). The same social group probably constructed both considering their virtually contemporaneous ceramic assemblages. These structures were built in loose sand that verged on collapsing during excavation (Figure Fifteen). In addition to these two structures, several sherds of Maverick Mountain series pottery were recovered from another late thirteenth century pithouse just to the west of the Daley site, preserved remarkably only a few centimeters below the gravel surface of First Avenue in Thatcher (Figure Sixteen). This pithouse had also been built in loose sand within a large channel.

The two pit structures excavated at the Daley site did not contain hearths or other thermal features, suggesting the occupations did not extend into the winter. Underlying and intrusive erosional channels in these features suggest that the area was an active alluvial setting before, during, and after the occupation. Hence, it is unlikely that these structures were occupied for more than a few months and would have been in peril during the summer monsoon thunderstorms or the winter rainy season. This occupation was, therefore, more of an encampment than a settlement.

The Maverick Mountain-dominated decorated ceramic assemblages from the pit structures can be contrasted with the large and varied decorated assemblage from the nearby trash-filled wash (Feature 7). Although contemporaneous with the pit structure component at the current level of temporal resolution, not a single example of Maverick Mountain pottery was identified in the 276 decorated sherds and partial vessels recovered from this unique deposit. Indeed, this deposit contained multiple examples of every other decorated ceramic type expected from a late thirteenth century context in the Safford Basin, including San Carlos Red-on-brown, Cibola White Ware, White Mountain Red Ware, MacDonald Decorated Corrugated, Casa Grande Red-on-buff, and early Zuni Glaze Ware. The deposit contained a large number of polishing stones and partially reconstructable vessels, some of which exhibited secondary burning. Ethnographic accounts document the use of large sherds and partially broken pots to protect vessels being fired from direct contact with the fuel (Rice 1987:154). The trash may have been generated by ceramic production activities that occurred at the margins of the main settlement.

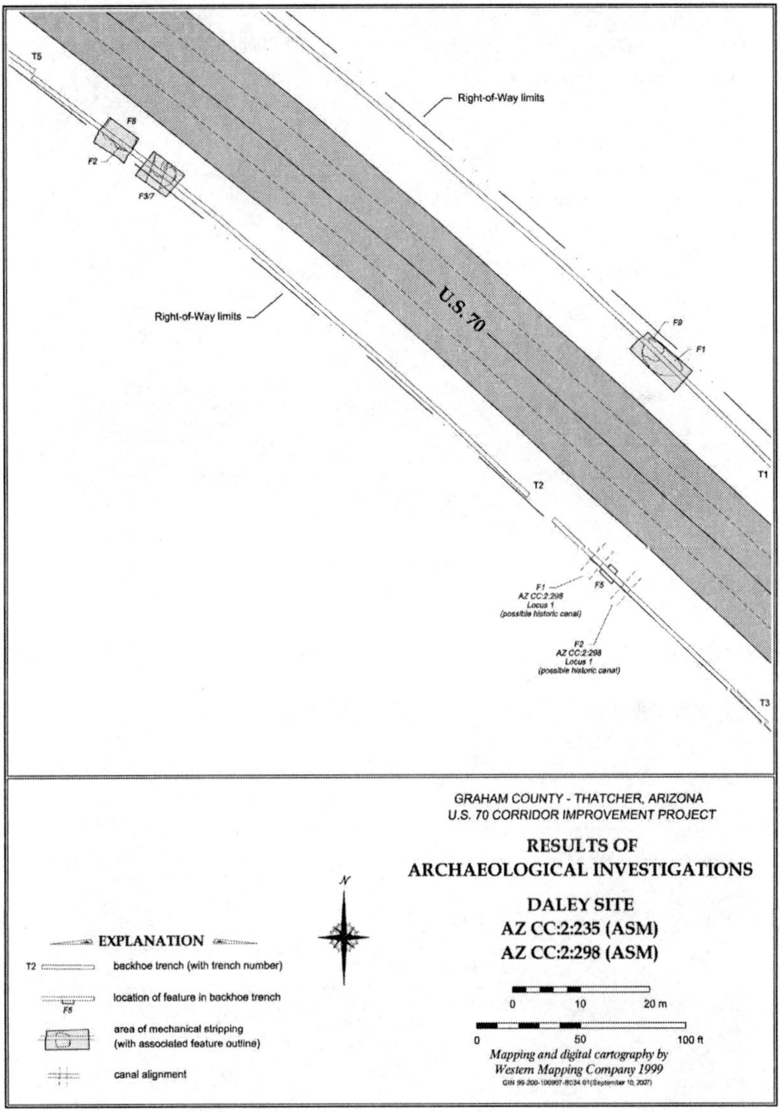


Figure 14: Location of test trenches and subsurface features at the Daley Site (AZ CC:2:235[ASM]).

Figure 15: Feature 9, a collapsed pitstructure in sandy substrate, at the Daley Site.

Figure 16: Plaster lined hearth exposed in $13^{th}$ century pithouse found beneath First Avenue in Thatcher, Arizona.

Maverick Mountain series pottery was also nearly absent from the compound and underlying component of the Daley site investigated by Eastern Arizona College. The decorated assemblage recovered from these excavations was dominated by Casa Grande Red-on-buff. The preponderance of buff ware and the presence of cremations suggest that the compound inhabitants had close ties to the Middle Gila Hohokam. However, corrugated wares were also present in relatively high frequency in both DAI's and EAC's excavations, suggesting influence from the Mogollon Highlands. The near-absence of Gila Polychrome from excavated contexts in both projects establishes an end date for these deposits in the early 1300s.

Although cautionary tales abound regarding the use of decorated pottery to define cultural groups, Maverick Mountain series pottery bears a strong resemblance to Kayenta and Tusayan decorated wares from northeastern Arizona and may indicate migration from these areas, especially when found in high frequency (Lyons and Lindsay 2006). Well-documented Kayenta/Tusayan enclaves at Point of Pines (Lindsay 1987), the Goat Hill site (Woodson 1995; 1999), and Reeve Ruin in the San Pedro Valley (Di Peso 1958) are associated with high densities of Maverick Mountain series pottery. In the case of the Daley site, dramatic differences in the distribution of Maverick Mountain Polychrome within a single site suggest co-residence by local groups in compounds and Kayenta/Tusayan migrants in a temporary camp at the edge of the settlement. Whether the migrants moved on or were ultimately accepted into the settlement cannot be determined given the available evidence.

## Conclusions

Woodson (1995, 1999), Lekson et al. (2002), and Neuzil (2006a) have focused attention on the late prehistoric migrations of Kayenta/Tusayan groups from northeastern Arizona into the Upper Gila Valley and Safford Basin. The findings from Roadrunner Village, Owens-Colvin, and the Daley site and other period settlements seem to indicate that this was not an isolated event or a unique interval in the region. Instead, it was a continuation of a pattern of cultural dynamics with great time depth in the region. Sayles (1945) viewed the "San Simon" Mogollon, the local inhabitants of the San Simon and upper Gila River valleys, as the descendants of Archaic hunter-gatherers and early farmers. Current research in the vicinity of Tucson and along the San Pedro River in southern Arizona substantiates this view (Gregory 2001; Heidke 1999; Huckell 1995; Mabry 1998, 1999). The early farmers of the Sonoran Desert and vicinity shared ceramic manufacturing techniques and designs, became increasingly dependent on irrigation agriculture, and as result built permanent villages centuries *before* the distinctive suite of traits researchers typically associate with the beginning of the Phoenix Basin Hohokam (i.e., buffwares, ballcourts, and formalized cremation burial and related burial goods) ca. A.D. 700 (Colek-Torrello 1995; Whittlesey 1995). While past scholars of the Safford Basin inferred that Phoenix Basin influence in this area was limited to the adoption of Hohokam material culture by local groups, a viable alternative explanation is that groups from the Middle Gila Valley immigrated into the region and coresided with local groups.

Prior to the Bylas phase, Roadrunner Village appears to have been inhabited primarily by local people. Some Hohokam-like traits become evident after A.D. 600, and include relatively large quantities of buff ware pottery, the advent of some Hohokam domestic architecture, and the appearance of some Hohokam religious paraphernalia. Although a substantial percentage of Hohokam Buffware pottery is typical of Safford Basin sites during the Eden phase and later, the co-occurrence of buffware pottery with indications of on-site paddle-and-anvil pottery production and the apparent adoption of Hohokam domestic architecture and religion is compelling. Domestic attributes such as house construction techniques and pottery manufacturing techniques are generally considered to be conservative and likely to persist with an individual as basic skills through their life. Likewise, the presence of items associated with the Hohokam cremation ritual, which was not practiced by the Mogollon, would seem to be associated with Hohokam individuals and the beliefs that they practice. The clustering of Hohokam domestic traits and religious paraphernalia at Roadrunner Village is viewed as indicative of the presence of people from the Hohokam realm west of the Safford Basin, just as the presence of "distinctive human and animal figurines, a palette, a pottery anvil, a censer fragment, and an assemblage of palette malleated ceramics" at Owens-Colvin Site, also in the Safford Basin, indicated Hohokam presence to Rule (1993:51). Marriage into a Mogollon village by Hohokam women is the most parsimonious explanation for this distribution of traits at Roadrunner Village, if domestic architecture and at least some aspects of ritual were female-dominated in Hohokam culture. Intermarriage may represent a crucial means by which ethnogenesis took place in the ancient Southwest. Likewise, the presence of Anasazi traits at Grasshopper (Reid and Whittlesey 2007) also seem best explained by intermarriage of migrants and locals.

If this interpretation of Roadrunner Village is accurate, and the usual caveats relating to incompletely investigated sites do apply in this case, then migration and cultural interaction of people with different culture backgrounds and perhaps ethnicities have great time depth in the Safford Basin. While genetic affiliation cannot be evaluated without a large and well-analyzed burial population (and may not be correlated with cultural patterns), material cultural differences in *how* people lived do show that different groups of people coresided in the Safford Basin much earlier than the Bylas phase, and the nature of their interactions has taken many different forms in this region.

By the Bylas phase, the local population in the Safford Basin was strongly influenced by Hohokam groups to the west and Mogollon groups to the north and east. The extent to which Hohokam, Mogollon, and other groups contributed to the identity of the late Formative period inhabitants of the Safford Basin requires much more data to be fully evaluated. For example, the Eden phase occupation of the Owens-Colvin site near Thatcher has been variously interpreted as "an actual Hohokam presence . . . as well as continued occupation by an 'indigenous' Mogollon population", apparently living together at this site (Rule 1993:51), and as "an essentially San Simon Branch Mogollon village that engaged heavily in trade with nearby Hohokam villages" (Jernigan 1993:5) However, varying mixes of "Hohokam" and "Mogollon" groups probably inhabited many of Bylas phase compounds in the region, including those associated with the primary occupation of the Daley site. The Bylas phase may represent, as Johnson and Wasley (1966) originally envisioned it, the crystallization of Hohokam/Mogollon interactions into something new, which they termed "Western Pueblo." The origins of this development must lie in the preceeding Eden phase, as they seem to do at Roadrunner Village.

Late in the Bylas phase, Kayenta/Tusayan migrants from northeastern Arizona reached the Point of Pines area and subsequently penetrated into the Safford Basin. The appearance of these migrants in southeastern Arizona meshes well with evidence for widespread abandonment of the Four Corners area in the late 1200s. Relatively large migrant groups built a few settlements in favorable areas on the edge of the basin. Whereas the excavation of the Goat Hill site seems to indicate that some immigrants from the Colorado Plateau lived, at least for a time, in segregated ethnic enclaves, the relationships between other migrants and local groups appears to have been much closer, as shown by the results of DAI's investigation of the Daley site, and Eastern Arizona College's investigations at Spear Ranch and Krider Kiva (Neuzil 2005b). Smaller, but probably more numerous, Kayenta/Tusayan migrant groups that could not form viable settlements camped on the margins of existing settlements to restock provisions for subsequent movements or to await acceptance into the settlement. Considering the high labor requirements of irrigation agriculture and the breadth of the Gila River floodplain, migrants may have been able to exchange labor in the fields for a share of the yield at least on a short-term basis. Ultimately, these groups would have either moved in or moved on. More Kayenta camps probably await discovery on edges of other large Bylas phase settlements in the region.

Thus far, considerable attention has been paid to Kayenta/Tusayan migrants. Although conspicuous and well-traveled, we should remember this group constitutes only a minority of the population, settling at the margins of settlements and on the edges of the Safford Basin, at least during the thirteenth century. To place these migrants in their proper socioeconomic context and to figure out what caused many of the inhabitants of the region to aggregate into large villages during the early 1400s we must also consider the local majority. In addition, recovery of early Zuni Glaze Wares at the Daley Site raises the possibility that the migrant or local population may have included Zuni groups or at least households that were interacting with these groups. Early Zuni Glaze Ware can be confused with other ceramics such as White Mountain Red Ware and it is likely that much more of the former is present in the Safford Basin than has been recognized to date.

It is both exciting and daunting to consider Safford Basin prehistory. The importance of the region in the prehistory of southern Arizona is irrefutable, despite decades of neglect. However, with the possible exception of the thirteenth century, we are still far from adequately characterizing a single interval within the sequence and excavated data are virtually absent for entire centuries in the sequence. Sites such as Daley, Owens-Colvin, and Roadrunner Village provide tantalizing clues about the interrelationships of Safford Basin people during earlier periods, preceding and leading up to the "Salado" horizon, and ultimately the ethnohistoric Pueblos. Although development and vandalism have heavily impacted much of the region, a considerable portion of the archaeological record remains intact beneath the fields. Understanding Safford Basin prehistory has great potential to inform us not only about the events that took place within the region, but also on the relationship between archaeological and living cultures and how the latter is transmitted and transformed in a dynamic multi-cultural setting.

As Reid and Whittlesey (2007) have pointed out, migration and multi-cultural communities were the rule, not the exception in the ancient Southwest. Few regions demonstrate this more convincingly than the Safford Basin.

# CHAPTER THREE

# PREHISTORIC GILA RIVER CANALS OF THE SAFFORD BASIN, SOUTHEASTERN ARIZONA: AN INITIAL CONSIDERATION

JAMES A. NEELY AND EVERETT J. MURPHY

To sustain subsistence requirements for food, fiber, and other products, the prehistoric inhabitants of the Safford Basin used agriculture (Clark 2004; Neely 2008a) to achieve a necessary crop yield through manipulations of plants and their physical environment (Bye and Shuster 1984:127). They were able to achieve their goal of sufficient crop production by using several environmental sub-zones, or microenvironments (Coe and Flannery 1964), and by manually modifying the landscape to receive additional sources of moisture as well as better retain and more efficiently distribute the moisture that was naturally available. Archaeological survey of the Safford Basin has provided evidence of a three-part prehistoric agricultural strategy (Neely 2008a). These are: (1) widespread dry-farming/runoff fields receiving moisture only from rainfall and snowfall; (2) fields in the foothills of the Pinaleño Mountains irrigated by canal systems taking water from springs and runoff from rainfall and snowmelt; and (3) fields on the floodplain irrigated by canal systems taking water from the Gila River.

It is highly probable that a great deal of effort was expended in the selection and modification of plants to achieve required crop yields. Unfortunately, as has been previously discussed more fully (Neely 2008a), at this time we do not have the evidence to do more than make a few general statements regarding this aspect of the overall agricultural diversification strategy (e.g., Diehl 2004; Neely 2008a; Neely and Doolittle 2004; Smith 2004). For the present, the reconstruction of agricultural strategies must depend most heavily on relic water management infrastructure and agricultural fields as well as archaeological sites and artifacts. Although other water sources were present and utilized (Neely 2008a), it is proposed that the well-planned excavation of several canals taking water from the Gila River supported a large portion of the basin's prehistoric crop production, and guaranteed its yearly availability.

The objectives of the present study are fourfold: (1) to investigate the evidence for the presence of prehistoric Gila River-source canals in the Safford Basin; (2) to evaluate the evidence for the prehistoric origins of certain historic Gila River canals; (3) to examine the data concerning when those canals were originally excavated and used; and (4) to consider the possibilities of who—which prehistoric group(s) or "culture(s)"—planned, excavated, and used those canals. We have found this to be a difficult set of goals to achieve by means of a study largely based on archaeological survey. This paper should be regarded as an initial consideration of these subjects. The findings should be considered tentative; undoubtedly they will be modified and refined as additional archaeological and historical work is accomplished. Some aspects of this paper are of a speculative or hypothetical nature, and some readers will be dismayed by this and think our effort to be a waste of time. But, if logical, well founded, and clearly labeled, such hypotheses are warranted when conclusive data are absent. Hopefully this study will serve as a starting point to guide future research, if for no other reason than to prove us wrong.

Specifically, this paper focuses on the canals with their heads or offtakes in the eastern portion of the Safford Basin, from just northeast of the small community of San José to just west of the town of Thatcher (Figure Seventeen). It is an expanded version of a section of a more encompassing article discussing the prehistoric agricultural strategies of the Safford Basin (Neely 2008a) recently submitted for publication. It also incorporates new information gleaned from additional fieldwork and research that augments our studies on agricultural systems conducted over the last several months. While a clear picture of the prehistoric fields and canal systems associated with the Gila River will probably never be achieved, several lines of evidence have permitted an initial consideration. As noted, the information utilized was largely obtained by survey, although a few small excavations have provided important and substantive data. Historical studies have also significantly contributed to this paper.

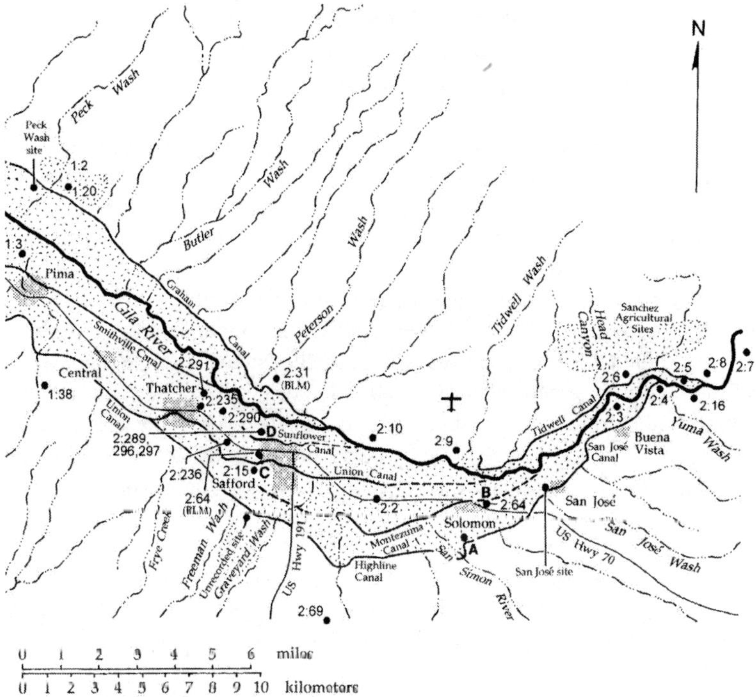


Figure 17: Map of a portion of the Safford Basin centering on the City of Safford. This map shows the location of sites mentioned in the text as well as the canals proposed to have prehistoric counterparts. The numerals (e.g., 2:290) represent the grid square and site number in the Arizona State Museum (ASM) site survey (AZSITE) system. The state and quadrangle designations (i.e., AZ and CC) have not been included due to lack of space. Note that sites 2:31 and 2:64 are from the Bureau of Land Management (BLM) site survey system files

With the information at hand, it appears that the Safford Basin was second in the American Southwest only to the Phoenix Basin in the intensity and extent of river-based canal use and the size of the related area of agriculturally utilized land in prehistoric times (Doolittle 2008; Neely 2008a). The surface evidence of prehistoric Gila River-based canals is rapidly diminishing while the sub-surface evidence is elusive and difficult to obtain, making prehistoric Gila River canal use one of the most difficult subjects in the study of agricultural development in the Safford Basin. This is due largely to the destruction and concealment of the canals and related fields by intensive historic/modern farming practices and the growth of communities paralleling the river. Paradoxically, in a few cases modern farming and public works have aided our study of ancient agriculture with excavations that disclosed prehistoric remains.

Archaeological evidence for the proposed prehistoric floodplain fields and canals is tantalizing, but lacking in quantity. A few excavations (Botsford and Kinkade 1993; Clark 2004, 2006; Huckleberry 2005; Lascaux and Huckleberry 2006) have provided important, but limited, information. When compared with the tangible evidence obtained primarily through survey in other parts of the Safford Basin (e.g., Doolittle and Neely 2004; Neely 2001a, 2005a, 2008a; Rinker 1998), the evidence for Gila River canals and related fields is sparse. However, the excavated evidence, when combined with settlement pattern data and historic information is sufficient to tentatively reconstruct the routes of some of the prehistoric canals. This reconstruction, augmented by Doolittle's (2008) study of the floodplain fields of the basin, presents a developing picture of Gila River floodplain agriculture.

The majority of the historic information (Colvin 1997, 1998; Colvin and Cook 2006; Fewkes 1898, 1904; Ramenofsky 1984; Southworth 1919) indicates canal-irrigated cultivation of the Gila River floodplain was being conducted mostly by Hispanic agriculturalists by 1870-1871. However, Spanish explorers may have known the area by the eighteenth century. Considering the 1757 arrival of the priest Bartolomé Sanchez in the Cliff, New Mexico/Upper Gila River area, some 100 kilometers east of the Safford Basin (Ackerly 1997; Doolittle 2000:387; Sanchez 1856), it seems quite possible that Hispanic agriculturalists occupied the Safford Basin earlier than 1870. A letter written by R. Louis Michelena (1990), indicating that Hispanic occupants of the Safford Basin were excavating canals as early as 1853, supports this possibility. This letter, and its importance, will be discussed below. Interestingly, like the early Hispanic and European agriculturalists in the Safford Basin, Padre Sanchez noted the presence of prehistoric agricultural irrigation canals in the Upper Gila River region (Ackerly 1997:354; Doolittle 2000:387; Sanchez 1856).

It is proposed that prehistoric canals were present on both sides of the Gila River in the Safford Basin (Figure Seventeen). The names of the canals used herein have been adopted from historic studies (Colvin 1997, 1998; Colvin and Cook 2006; Ramenofsky 1984), early archaeological studies (Fewkes 1898, 1904), and the United States Geological Service (USGS)

7.5-minute topographic maps of this area. A few of the historic canals in the Safford Basin are herein proposed to be refurbishments of prehistoric canals or closely follow the paths of prehistoric antecedents. For the time being, we must assume that the remaining canals are of historic origin.

## Prehistoric Canals South of the Gila River

As of 2006, evidence supports the presence of at least five prehistoric canals irrigating fields on the south side of the river. From east to west, they are historically named the Old San José-Fourness Canal, the San José-Highline Canal, the Montezuma Canal, the Union Canal, and the Sunflower Canal (Figures Seventeen and Eighteen).

### The Old San José-Fourness Canal

Named the "Old San José Canal" by Babb (1901:Figure 201) and the "Fourness Canal" by Southworth (1919:163), this is the easternmost of the canals that may have had a prehistoric origin. The early historic, and possible prehistoric, head of this canal, as explained by Southworth (1919:163), served both the San José Canal and the Fourness Canal prior to 1891, when the canals were separated and the head of the San José Canal was moved a short distance down-river.

Today, all that is apparently preserved of this canal is a small segment of its channel visible approximately 1.8 kilometers south-southwest of the present head of the San José Canal, as close as 50 meters to its east, and immediately east of the present paved Buena Vista Road (see point "A" on Figure Eighteen). This canal courses just above the toe of a high terrace (Figure Nineteen), is higher than the San José Canal at this point, and in places has a conspicuous channel and well-defined outer (western) berm

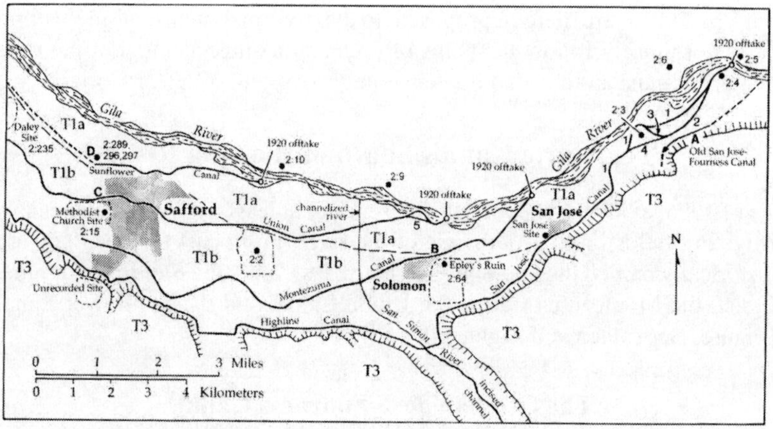


Figure 18: Map of the eastern portion of the Safford Basin. The canals proposed to have prehistoric counterparts are shown in their 1920 map (Arizona State Water Commissioner 1920) locations relative to the terraces south of the Gila River (Doolittle 2008; Huckleberry 2005:Figure 3.2) as well as the major archaeological sites and modern communities of the area. Numbered San José Canal segments: (1) pre-1897; (2) post-1897; (3) post 1897 and pre-1920; (4) later than (3) but pre-1920. Numbered Union Canal segment: (5) post 1920 and pre-1935. A dashed line indicates an approximate route or boundary. U.S. Highway 70 generally follows the northern boundary of the T1b terrace.

(Figure Twenty). It is about 1.5 m wide at its present floor level, and is about 50 cm deep, making it substantially smaller than the present channel of the San José Canal (discussed below). The flow in this canal would have been augmented by runoff from the high terrace immediately to the east. This canal was traced northward to a prominent "nose" of the high terrace but it is uncertain, due to road construction and alluviation, whether this canal followed the contours around the nose of the terrace or if it coursed outward and northeastward. Because of the topography, it is clear that this canal did not branch from the San José Canal but had its head at the Gila River further upstream. Neither the earliest detailed map (Arizona State Water Commissioner [ASWC] 1920) nor the earliest aerial photos (U. S. Department of the Interior [USDI] 1935:A20 04009 273-8) we have found of this area revealed any clear indications of the upper portion or head of this canal, although it is conceivable that the channel may have followed the well-considered northeastward route we have projected on Figure Eighteen.

Figure 19: A wide-angle photograph of the terrace face into which has been constructed the small canal also shown in Figure Twenty. This is probably a segment of the Old San José - Fourness Canal. The white colored linear feature seen in the center of the photograph is the lower (western) berm of the canal. This small canal generally parallels, and lies as close as 50 meters east of, the San José Canal near the Buena Vista Site. Looking south.

Figure 20: The small canal channel shown in Figure Nineteen, and believed to be a segment of the Old San José - Fourness Canal. The six-foot (1.8 meter) tall man is standing in the center of the channel. Looking southwest.

We then traced this canal to the west-southwest some 900 meters, to a point at which the well-defined channel turns southeastward to enter a large drainage bisecting the terrace (Figure Eighteen). Unfortunately, we could not trace it further because it was on posted private land. Aerial photos (USDI 1935:A20 04009 273-8) verify that the canal did turn southeastward into the drainage arroyo, but we could not determine if the canal was directed into the arroyo to irrigate the alluvial fan at its mouth, or if it crossed the fan and continued southward. In the 900 meters we traced this canal it was seen to descend (from north to south) approximately 3.0 meters. This drop, although not measured from the original floor of the canal, suggests a gentle grade of about 0.3 percent.

Babb (1901:341, Figure 201) briefly mentions this small canal as functioning in 1899, and Southworth (1919:163) and Colvin (1997, 1998) discuss it in only slightly greater detail, but we have failed to find further mention of it in any other historic studies or documents. While we have no tangible proof, it is entirely possible that this feature represents a prehistoric canal that may have seen historic refurbishment. We base this hypothesis on the proximity of two or three prehistoric habitation sites (i.e., AZ CC:2:16 and AZ CC:2:4[ASM]), and perhaps AZ CC:2:3(ASM), all with estimated overlapping times of occupation (Neuzil 2005b; Arizona State Museum Site Survey files) on the south side of the Gila River, and probably two contemporaneously occupied sites AZ CC:2:8, and 2:5(ASM) on the north side of the river. This is admittedly one of the most circumstantial of the historic canals we propose to have possibly been a rejuvenated prehistoric canal or to have a possible prehistoric counterpart. It illustrates the pattern of having habitation sites with contemporaneous or overlapping occupations situated along its channel, but lacks some of the other supporting evidence we have found with other canals in the basin.

## The San José-Highline Canal

The next canal down-river, the San José Canal (Figures Seventeen and Eighteen), is one of the canals mentioned in historic studies noting the refurbishment of prehistoric canals by the early Hispanic occupants of the Safford Basin (Colvin 1997, 1998; Ramenofsky 1984). J. Walter Fewkes, the first professional archaeologist to excavate in the Safford Basin, noted the presence of this canal, an abundance of abandoned prehistoric canals, and that: ". . . sections of the modern ditches follow the course of the ancient waterways" (Fewkes 1898:613). Fewkes (1904:178) specifically states that "an old settler" in Solomonville told him that the "modern" San José ditch follows, in part, the course of an ancient canal.

Our survey indicates that most of the present channel of this canal probably closely approximates its original course, but we were not able to locate the earlier head of this canal that Southworth (1919:163) states was a sort distance up-stream until 1891. The prehistoric existence of this canal is supported by four observations: (1) the presence of remnants of old, apparently prehistoric, segments of the canal, (2) the presence of remnants of apparently prehistoric offtakes (subsidiary, lateral, or secondary canals) from the canal, (3) the occurrence of archaeological sites with overlapping occupations paralleling the course of the canal, and (4) historic information that this canal was ancient and had been refurbished by 19th century Hispanic and Anglo agriculturalists.

As with other canals to be discussed, the dimensions (i.e., width, depth, and length) of the prehistoric counterpart of the San José Canal are not known. Although Fewkes (1904:175) comments on the apparently large size of the prehistoric irrigation ditches in the Safford Basin, Southworth (1919:162) notes that the Brown Canal (i.e., the Michelena-Tidwell Canal-discussed below) was only three feet wide until 1883. Today, the San José Canal is quite large (Figures Seventeen, Eighteen, and Twenty One), and it is feasible that its present day dimensions approximate those of the prehistoric canal as well. The scale of a canal system reflects the value (i.e., real and perceived need) placed on it by its agriculturally based users, and examples may be seen in the large-scale works found through archaeology. The large Hohokam canals of the Phoenix Basin (Abbott et al. 2006; Doolittle 2000:347-409; Hunt et al. 2005) are prime examples. Crown (1984, 1987:Figures 2, 3; also see Doolittle 2000:390-392) reports equally large prehistoric canals located near Florence, Arizona that head about 145-airline kilometers west-northwest of the city of Safford. One of those canals, also on the south bank of the Gila River, evidently once serviced the area of the Casa Grande Ruin. The mind boggles at the amount of time and effort expended, in some cases for apparently little return, on some of the massive prehistoric water management features and systems in Mexico (e.g., the canals/aqueducts on the Cerro Tetzcotzingo and Cerro Purificación in the Texcoco area [Doolittle 1990:127-135] or the Purron Dam in southern Puebla [Neely 2005b; Neely et al. 2005; Woodbury and Neely 1972]).

Figure 21: Photograph of the San José Canal a short distance down-canal from its head at the Gila River. Note the six-foot (1.8 meter) tall man standing on the canal's right bank for scale. Looking southwest.

Our 2006 survey provided two scenarios for the relationship of the San José Canal and the Highline Canal. The first of these is that the San José Canal terminated in fields at a point about 950 meters to the northeast of the Highline Canal (see point "A" on Figure Seventeen), and that its connection with the Highline Canal represents a historic extension. Babb's (1901:341:Figure 231) map apparently supports this scenario by showing the 1898 course of the San José Canal ending near Solomonville. However, Babb's map is of a very small scale and displays a number of apparent discrepancies in canal routes and names. The apparent route of the Highline Canal, which Babb terms the "Enterprise Canal", is shown as "under Construction." In addition, from this source there is no way to determine if the Highline Canal was to follow the relict channel of a prehistoric canal.

The second scenario is that the present connection of the San José Canal and the Highline Canal existed in prehistoric times, making the total length of this canal (i.e., the combined San José Canal and the Highline Canal) approximately 25 kilometers. At this time, we propose that the second scenario is not only feasible but also quite likely.

*The Relationship of the San José Canal with the Highline Canal.* The two most likely objections that may be raised to our proposal for the prehistoric existence of the combined San José Canal-Highline Canal, and its feasibility in general, are: (1) the overall length of the combined canal, and (2) the north-south oriented natural drainages (e.g., the San Simon River) that would have stood in the way of making this one long canal.

The first of these two objections is easily countered. The approximate 25-kilometer length of the San José-Highline Canal is less than the over 30 kilometer long prehistoric canal reported by Crown (1984, 1987; also see Doolittle 2000:390-392) that serviced the Casa Grande Ruin area. Longer canals are also found in other locations in the Hohokam area (Abbott et al. 2006; Doolittle 2000:347-409). The San José-Highline Canal functions well today, so there is no real reason why it could not have functioned prehistorically.

Perhaps the major objection that may be raised to our proposal is that the postulated ancient canal would have had to be constructed to cross north-south trending drainages, including the San Simon River channel, in order to function. While today this canal crosses north-south drainages by means of siphons, such crossings may be seen as considerable barriers for prehistoric engineers. As a result, ancient counterparts of the historic canals have been generally dismissed as technologically impossible or impracticable in prehistoric times.

Evidence for the engineering knowledge and technological capability to overcome this problem has been archaeologically documented in the Safford Basin and elsewhere in the American Southwest. The discovery of the prehistoric raised nearly 80 meter long aqueduct south of the city of Safford in Marijilda Canyon (Neely 2008a, 2008b; Neely and Crary 1998) indicates that crossing such drainages was not an insurmountable obstacle for the prehistoric agriculturalists. By having the knowledge to construct this aqueduct, it is entirely reasonable that the prehistoric agriculturalists could have constructed a structure to span large drainages. This structure probably took the form of a canoa (*cañoa*, Sp.-cf. Doolittle 2000:Figure 10.4), such as that seen spanning a wide drainage today in the vicinity of Las Trampas, New Mexico (Figures Twenty Two and Twenty Three).

Figure 22: A canoa (*cañoa*) spanning an arroyo just east of the town of Trampas, New Mexico. Note the men standing at the left (north) side of the photograph for scale. Looking east.

Figure 23: Another view of the Trampas, New Mexico canoa, looking south. Two long logs have been cut to form U-shaped channels and then partially overlapped to span the arroyo.

The Mullen Wash aqueduct, located in the Verde Basin of Arizona, provides an additional excellent example that reinforces the probable use of aqueducts in the form of canoas. Mindeleff (1896) first recorded the limestone canal of which the Mullen Wash aqueduct was a part. It was better documented and more thoroughly studied later by Midvale (1946). Doolittle (2000:377, Fig. 10.13) observes that the break in the Mullen Wash aqueduct, where it crosses the active channel of Mullen Wash, was probably spanned by " . . . split and hollowed logs . . . " (Doolittle 2000:Fig. 10.4; Van West and Altschul 1997:346, Plate 9.4).

At the present juncture of the San José Canal with the Highline Canal, the San Simon River channel ranges from roughly 19 to 26 meters across, depending on the place on the constantly changing banks one takes a measurement. It is currently about 3 to 3.5 meters from ground level to the bed of the river (measured when dry). However, Gary Huckleberry (personal communication 2006) has informed us that the current depth and width of this entrenchment occurred after about 1900, and that the San Simon channel was previously less wide and deep (Olmstead 1919). To span the river the prehistoric agriculturalists could have constructed an aqueduct constructed of several large trees with U-shaped channels cut into them (Figure Twenty Three). Several such large logs, laid side-by-side and overlapping end-to-end, and supported by a framework of vertical logs imbedded in the river bottom and banks (see Figures Twenty Two and Twenty Three), could easily have spanned the width of the San Simon River and connected the ancient segments of the large San José and Highline Canals. Near the small village of San Juan del Maguey, in the state of Jalisco in Highland Central Mexico, Neely (personal observation 1970) has also seen a similar task accomplished with large agave leaves. In this case, many large agave leaves had been overlapped side-by-side and end-to-end, and sewn together with agave fiber, to span a wide drainage. This very strong and flexible conduit was supported on a framework of logs.

While indisputable proof of the course and proposed extension of the San José Canal to include the Highline Canal is not now available; there is enough evidence present to at least propose it as a credible possibility. See the Observations section, below, for additional discussion on this matter.

*The Buena Vista Segment of the San José-Highline Canal.* About 2.7 kilometers southwest of the present head of the San José Canal is located the Curtis Ranch, on which the large Buena Vista (Curtis) site (AZ CC:2:3[ASM]-Fewkes 1898, 1904; Mills and Mills 1978) is situated. The ranch and site are located to the west of the present channel of the San José Canal (Figures Seventeen and Eighteen). With permission from the landowner, Mr. Brooks Curtis, we investigated the remnants of the prehistoric canal that followed the perimeter of the peninsular-like terrace upon which the Buena Vista site was constructed. Unfortunately, much of this canal has been bulldozed inward toward the center of the terrace. This was done within about the last 15 years to increase the area of the modern fields located primarily to the north and west of the terrace. However, we were able to find a well-defined segment of the canal below a historic dwelling situated a short distance east of the "nose" of the terrace. About three meters below and approximately 15 meters to the north of this dwelling, this canal segment was concealed in dense brush. The channel was situated about half way up the terrace face. The upper portion of the terrace appears to have been cut back to form a ledge into which the canal was excavated. The present 60 cm deep channel is about 3.5 meters wide at its bottom. Its outer (north) berm was clearly visible and the berm's interior face suggested a canal with nearly vertical sides.

J. W. Fewkes was apparently the first to record the canal channel at the perimeter of the terrace on which the Buena Vista site is situated. In viewing Fewkes' (1898:Plate XVI; 1904:Plate LXVI) map of the Buena Vista site (see Figure Forty One), we noticed that it shows a canal, flowing from a more northwesterly route than that presently taken by the San José Canal, met the "nose" of the terrace and then followed its western edge toward the south. This is inconsistent with our findings that the canal followed the entire perimeter of the terrace, and is also inconsistent with the situation as illustrated by an early map (ASWC 1920) and slightly later aerial photos (USDI 1935:A20 04009 273-8). These findings brought to mind two questions: (1) was the segment of the canal we describe above a part of the prehistoric primary canal; and (2) did people from the community of San José and/or the Curtis family, who purchased the land in 1903 (Brooks Curtis, personal communication 2006), move the northern segment of the San José Canal to its present location after Fewkes' map was made? We have no definite solution for this inconsistency; however, we envision two possible explanatory scenarios. The first is that the original primary canal was situated further to the west than the San José Canal is presently located, and sometime after 1897 and before 1920 the northern segment of the canal was relocated (Figure Eighteen). The second scenario, and probably the less likely, is that the cartographer mapping the site for Fewkes in 1897 may have erred by showing a still extant secondary canal (now named the Fourness Canal [Colvin 1998:21-22; Neely, personal observation 2006]) flowing into the terrace canal at this point as the primary source canal. If the latter scenario was the case, the cartographer apparently compounded this error when he failed to illustrate the eastern half of the true primary canal (i.e., the prehistoric counterpart of the San José Canal) that followed the approximate 180-degree arc of the terrace edge (Figure Eighteen).

Babb's (1901:Figure 201) map of 1899 lacks the detail and scale to contribute. However, the 1920 map noted above designates the large bend of the canal that passed around the Buena Vista site's terrace as the “Old San José Canal”, and that it was “abandoned.” If the northern segment of the original prehistoric primary canal was relocated after 1897, perhaps due to its destruction during disastrous floods such as those of 1905-1906 and/or 1914 (Talley 1970:6, 9), it is possible that the eastern portion of the Buena Vista Canal seen on the 1920 map and 1935 aerial photos was part of that San José Canal realignment. In the early stages of our survey, we thought that the Buena Vista Canal might have been a secondary canal branching from the San José Canal at about the present location of a modern canal offtake that irrigates the fields to the west and southwest of the Buena Vista Site. However, in 2006, Mr. Brooks Curtis revealed that the canal that had bordered the Buena Vista site was not a secondary branch, but was once part of the primary San José Canal. Mr. Brooks recalled that his grandfather, father, and several other men used a “power shovel” to excavate a more direct channel for the San Jose Canal in the early 1900s, and thereby cut off the large bend that bordered the Buena Vista site terrace (Figure Eighteen). Therefore, as best we can determine at this time, the original course of the San José Canal was correctly shown by Fewkes' map. Then, sometime after 1897 but before 1920, the northern part of the canal's course was moved eastward and the channel was cut around the north side of the Buena Vista site terrace. Sometime later, but before 1920, the bend in the canal then present around the Buena Vista site terrace was cut off to straighten the channel. Obviously, written records of these matters would do much in reconstructing the history of the San José Canal. Unfortunately, such records have not been found, and the historic information available does not contribute to the solution of the many problems seen or the questions raised (see the “Observations” section, below).

The ruins of an adobe house (Figure Twenty Four) were found about 60 meters west-northwest of the point where the proposed post-1897 realignment of the San José Canal turned westward to go around the Curtis ranch and the Buena Vista site. We learned that a Señor Sixto Molina homesteaded the property on which this adobe house is situated in 1896

(Graham County Records Office, Homestead Index, Homestead Book No. 1, Page 157). As we traced the course of the San José Canal southward, we found many adobe house ruins located on both sides of the San José Canal between this point and the community of Solomon. Although we have not as yet attempted to research these structures, the locations of these ruins suggests that the present course of the San José Canal from this point southward has not changed since agriculturalists of the late 1800s and early 1900s, like their prehistoric counterparts, used the San José Canal and constructed their homes near the canal. The presence of a now defunct "hot" spring situated immediately east of the adobe house (Brooks Curtis, personal communication 2006) may explain the placement of this house. However, to further complicate the question of the realignment of the San José Canal, its location might also be used to suggest that the San José Canal was in its present location prior to the drawing of Fewkes' map. A white obelisk-shaped marble monument, erected in 1942, located immediately west of the adobe house (see Figure Twenty Four) commemorates the birth and death of four children from the Molina family. Assuming the children were buried at this location, the dates on the monument would indicate that the first child to die was buried here in 1899. This suggests, but by no means proves, that the adobe house was constructed here between 1896 and 1899.

*Offtakes/Branchings from, the San José-Highline Canal.* A possible prehistoric offtake or branching from the canal that passes around the Buena Vista site was found that courses directly west toward an area of presently cultivated fields. From this point the larger canal continues its course of about 10 degrees E of S, to complete its loop around the Buena Vista site. At this location, the larger canal was about 1.5 meters in depth, approximately four meters wide at ground level, and about 1.5 meters wide at its present floor. The smaller canal was about 1.5 meters deep, some two meters wide at ground level, and about one meter wide at its present floor.

Figure 24: The ruined adobe building found just west of the San José Canal at the northeastern edge of the terrace on which the Buena Vista site is located. This is one of the better preserved of many similar structures constructed on both sides of the San José Canal between the Buena Vista site and the Solomon (Solomonville) community.

The indistinct remnants of several very old, possibly prehistoric, offtakes were also found branching northwestward from what is now the Highline Canal. These secondary canals were about two meters wide at ground level and approximately 50 cm in depth. This segment of the Highline Canal courses just above the toe of a steep-faced high terrace (the T3 terrace-Figure Eighteen; Huckleberry 2005:Figure 3.2) that supports dense riparian vegetation. The area to the north of this canal is heavily cultivated today, as it probably was in prehistory.

*Sites Probably Associated with the Ancient Predecessor of the San José-Highline Canal.* From east to west, the five to eight habitation sites (Figure Seventeen) probably associated with the San José Canal are: the Yuma Wash site (AZ CC:2:16[ASM]-Brown 1973); AZ CC:2:4 ([ASM]-Arizona State Museum survey files); the Buena Vista (Curtis) site (AZ CC:2:3[ASM]-Brown 1973; Fewkes 1898, 1904; Mills and Mills 1978;

Tuohy 1960); the San José site (Fewkes 1898:614, 1904:173); and Epley's Ruin (AZ CC:2:64[ASM]-Fewkes 1898, 1904; Lascaux and Huckleberry 2006; Lascaux and Montgomery 2005). Sites AZ CC:2:2[ASM] and AZ CC:2:69(ASM) may well have been associated with the Highline segment of the San José Canal. In addition, the unnamed and unrecorded site that is now largely covered by apartments and a small shopping center on the rise of land located between South 17th and South 20th Avenues, and other sites in its vicinity, may also have been associated. Most of these sites are documented in the reports cited above, the Archaeological Site Survey files of the Arizona State Museum, as well as by Neely (2005a, 2008a); and Neuzil (2005b). As indicated by the reports of Bandelier (1892) and Fewkes (1898,1904), they represent the few sites that have, at least in part, escaped historic cultivation and urban construction. Fewke's (1898:610-616, 1904:170-175) observations on the large number of site remains present in the Solomonville and San José areas in the late 1800s are particularly insightful when considering prehistoric site and population densities and related subsistence requirements.

## The Montezuma Canal

The Montezuma Canal is believed to represent the next prehistoric canal to occur down-stream on the south side of the Gila River (Figures Seventeen and Eighteen). Historic information (Colvin 1998; Fewkes 1898:613) attests to this canal's prehistoric origins. Fewkes (1904:178) specifically states that "an old settler" in Solomonville (name changed to "Solomon" around 1900) told him that the "modern" Montezuma ditch follows, in part, the course of an ancient canal. In addition, like the Union and Sunflower Canals (below), this canal has nearby archaeological evidence that supports its attribution as a refurbished prehistoric canal and/or closely follows the course of a prehistoric antecedent. Its very name, "Montezuma", suggests its prehistoric origin, as demonstrated elsewhere (e.g., Aztec, Montezuma Well, etc.) where historic groups have applied such names to ancient features. Recent excavations, at the northeastern edge of Solomon (Solomonville) (see "B" on Figures Seventeen and Eighteen) by Tierra Right-of-Way Services, Inc., found segments of three prehistoric canals and one historic canal with east-west orientations near the functioning Montezuma Canal (Huckleberry 2005; Lascaux and Huckleberry 2006). The earliest of the prehistoric canals has been radiocarbon dated to cal. 190 B.C.-A.D. 10, making it the earliest dated canal in the Safford Basin and one of the earliest dated canals in the American Southwest. As noted in the "Observations", below, these dates augment the growing suite of dates from southern Arizona that indicate great antiquity for canal irrigation in the American Southwest.

Near the northeast edge of Solomon, the current Montezuma Canal turns due south and courses about 1.35 kilometers to where it branches from the San José Canal. However, the general topography of the area, as well as east-west orientations of the prehistoric canals found by Annick Lascaux and Gary Huckleberry (2006) near the northeast edge of the Epley's Ruin (AZ CC:2:64[ASM]) and the Solomonville community, seem to indicate that the head of the original Montezuma Canal was to the northeast. Unfortunately, our survey was unable to find any evidence of an offtake from the river, nor of the 2.5 to 3 kilometers of the upper portion of this canal that presumably led to the presently visible remnant of the canal. Our thoughts were that the original head or offtake of the Montezuma Canal from the Gila River was probably located about 1.3 kilometers north-northwest of the small historic community of San José, and approximately 6.5 kilometers down stream from the head of the San José Canal (see Figures Seventeen and Eighteen; Colvin and Cook 2006:11).

During our attempt to trace the upper portion of the Montezuma Canal by pedestrian survey, the only relatively undisturbed area in the vicinity was directly north of the Lascaux and Huckleberry excavations. In this area we did not find any evidence of a canal, but we did find the remains of a ruined adobe building in a dense stand of mesquite trees. This ruined adobe was reminiscent of other similar structures found to parallel the San José Canal, and was thought to support the theory that the head of the Montezuma Canal lay to the northeast. The 1920 map (ASWC 1920), found at the Gila Valley Irrigation District Office in Safford, resolved our quandary regarding the location of the probable original head of the Montezuma Canal. That map clearly shows the offtake of the Montezuma Canal was located very near where we had predicted it should be (Figures Seventeen and Eighteen). The map also shows the route of the canal channel to where it joined the now existing portion of the Montezuma Canal, about 450 meters west-northwest of the Lascaux and Huckleberry excavations (see point "B" on Figures Seventeen and Eighteen). Relative to the channel of the Montezuma Canal shown on the 1920 map, we now see that the ruined adobe noted above was situated just south of that primary channel, and thus maintained the pattern of early adobe ruins found paralleling the San José Canal. As the early aerial photos (USDI 1935:A20 04009 273-8) do not show the head or this portion of the Montezuma Canal, it is evident that its offtake was changed to branch from the San José Canal sometime between 1920 and 1935.

Considering the orientations and relatively small size of the Epley's Ruin prehistoric canals, Huckleberry (2005) comments that it is likely that the smallest of the three canals was a tertiary (lateral or field) canal, while the two others were secondary (distribution) canals and were probably branchings from an as yet to be determined primary canal. As noted above, the 1920 map shows the presently missing portion of the Montezuma Canal to course north of the Lascaux and Huckleberry excavations. Thus, due to the location of the Montezuma Canal, and because the topography slopes toward the river (north), it is unlikely that the canals found by Lascaux and Huckleberry were branches from its prehistoric counterpart. The exact location for the offtake for the Epley's Ruin canals remains, for now, unknown. However, it is conceivable that the prehistoric canal segments found by Lascaux and Huckleberry may have branched from the prehistoric counterpart of the San José Canal a short distance southwest of the San José site. From Fewkes' (1898:612-613, 1904:178-179) observations, the evidence for prehistoric canal irrigation was abundant and clearly defined in the area between San José and Solomonville in 1897. It is a shame that it is not so today.

We were unable to investigate the Montezuma Canal immediately west of the Lascaux and Huckleberry excavations due to heavy historic disturbances and constructions. However, the topography suggests that the present course of this canal west of Solomon probably closely parallels its original course. Today the canal terminates in an open field area at a distance of about seven kilometers west of the Lascaux and Huckleberry excavations, however, the 1920 map indicates that the Montezuma Canal extended further westward into the city of Safford (Figure Eighteen) to a point about halfway between the San José-Highline Canal and the Union Canal and about 500 meters south of the postulated southern limits of the Methodist Church Site (AZ CC:2:15[ASM]), discussed below. Babb's (1901:Figure 201) map suggests that in 1899 the Montezuma Canal extended to just southwest of Thatcher (Figure Eighteen).

*Sites Probably Associated with the Ancient Predecessor of the Montezuma Canal*. The habitation sites probably associated with the predecessor of the Montezuma Canal are (from east to west): the San José site (Fewkes 1898: 614, 1904:173); Epley's Ruin (AZ CC:2:64[ASM]-Fewkes 1898, 1904; Lascaux and Huckleberry 2006; Lascaux and Montgomery 2005); AZ CC:2:2(ASM); probably the Methodist Church site (AZ CC:2:15 [ASM]-Brown 1973; Crary 1997); and other sites now destroyed or buried.

## The Union Canal

The Union Canal appears to be the current counterpart of a fourth prehistoric canal located a bit further down-river on the south side of the Gila River (Figures Seventeen and Eighteen). The 1985 USGS *Safford, Ariz.* 7.5-minute Topographic quadrangle map and our survey indicate that this canal receives waters from a modern "aqueduct" (i.e., a buried pipeline) branching from the Highline Canal, and possibly from pumped wells. We found that east of South Barney Lane (where the canal appears to turn south) surface indications of the Union Canal have been obliterated by agricultural landform modifications and intensive cultivation.

An earlier head, perhaps the original head, of this canal took water from the Gila River at a point in the approximate center of Section 18, Township 7 South, Range 27 East (Figures Seventeen and Eighteen), approximately four and a half to five kilometers east of where the "aqueduct" is shown to join the surface channel (1985 USGS 7.5-minute topographic Safford Quadrangle). The location of this offtake is documented by interviews (e.g., Frank Quinn [a local Safford historian], personal communication 1997), and from older maps (e.g., Bureau of Land Management [BLM] 1973). The 1920 map noted above also indicates the head of the Union Canal to be located at nearly the exact point shown in Figures Seventeen and Eighteen, and by Colvin and Cook (2006:11). Attesting to the rapid human modifications affecting canal channels, the aerial photos (USDI 1935:A20 04009 273-8) indicate the channel was straightened just west of the offtake sometime between 1920 and 1935 (see "5" on Figure Eighteen). As was accomplished for the San José Canal at the location of the Buena Vista site, this straightening cut off a large bend in the channel of the Union Canal. To the west, the course of the historic canal, from just northeast of site AZ CC:2:2(ASM) (Figures Seventeen and Eighteen) to where it enters the current community of Thatcher, appears to be in about the same location as may be projected for its prehistoric predecessor. This observation is based on three factors. (1) Although we could not walk the course of the canal through much of the heart of Safford and Thatcher, the topography of the landscape into which the canal was excavated presents few alternatives in its route. (2) The presence of archaeologically documented remnants of prehistoric canals found nearby the present functioning canal. (3) Archaeological sites paralleling the course of the canal.

Two complexes of canals have been archaeologically documented just north of the present channel of the Union Canal (see point "C" on Figures

Seventeen and Eighteen). It must be noted that these canals were apparently located at or near the north-central edge of the Methodist Church site (AZ CC:2:15[ASM]).

The northernmost complex of canals (Figure Twenty Five) was found to contain three small prehistoric canal segments associated with prehistoric architecture. This complex of canals was found as a result of professional test excavations anticipating the construction of the present BLM office building in Safford (Botsford and Kinkade 1993). The site was named the BLM site and designated AZ CC:2:64(BLM), although it probably is part of the Methodist Church Site (AZ CC:2:15[ASM]). The site and canals were dated to ca. A.D. 850-1200 by means of associated ceramics samples.

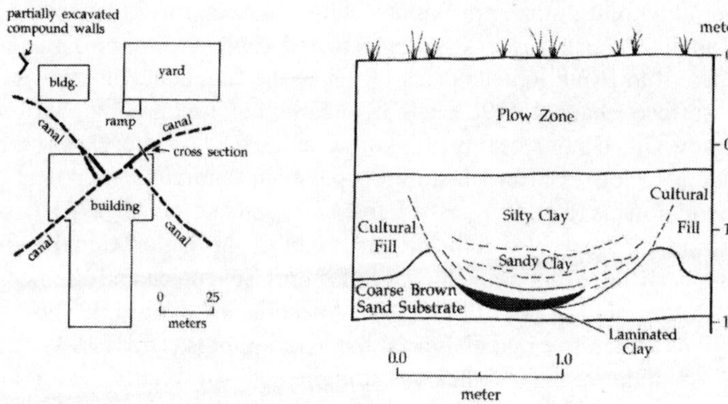


Figure 25: Drawings of the small prehistoric canals found during archaeological testing at the BLM Site (AZ CC:2:64[BLM] - see "C" on Figures Seventeen and Eighteen). The top of the plan drawing is north, and the cross-section is facing northeast. (Adapted from Botsford and Kinkade 1993).

The southernmost of these two complexes of canals (Figure Twenty Six), also in association with prehistoric architecture, was fortuitously found to the south about a year later. The owner and developer of the property excavated this site. Fortunately, Gay Kinkade (BLM Archaeologist), Everett Murphy (Arizona Archaeological Site Steward), and Joseph Crary witnessed this excavation and were able to briefly study the findings. The canal recorded by Crary (1997) was just north of and generally paralleled the present channel of the Union Canal. Crary was able to note that the route of this relatively large prehistoric channel had been modified several times

(Figure Twenty Six), possibly due to a "fine-tuning" of the channel grade for more efficient flow, or to repair washouts.

Following the Union Canal through Safford and Thatcher was quite difficult as it passes through developed neighborhoods, but we were able to see it in enough places to get a good idea of its topographic surroundings. Lack of time prevented us from tracing this canal more than about a kilometer west of Thatcher, just past where the Highline Canal presently joins the Union Canal.

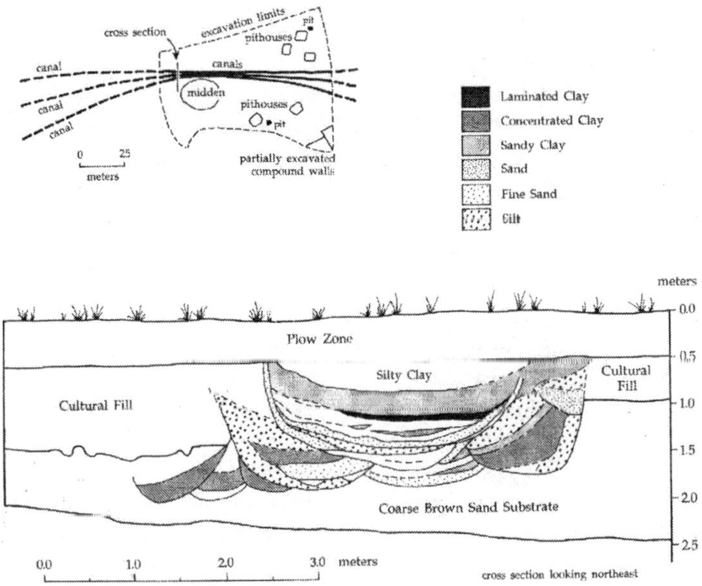


Figure 26: Drawings of the prehistoric canal found south of the BLM building in Safford, and just north of the functioning Union Canal. This canal may be a primary canal ancestral to the present Union Canal (see "C" on Figures Seventeen and Eighteen). The top of the plan drawing is north, and the cross-section is looking west. (Adapted from Crary 1997).

*Sites Probably Associated with the Ancient Predecessor of the Union Canal.* The proposed ancient predecessor of the Union Canal apparently had as many as seven habitation sites in association. These sites (from east to west) are: Epley's Ruin (AZ CC:2:64[ASM]); AZ CC:2:2(ASM); the Methodist

Church site (AZ CC:2:15[ASM]-Brown 1973; Crary 1997); the BLM site (AZ CC:2:64[BLM]-Botsford and Kinkade 1993); sites AZ CC:2:290(ASM), and AZ CC:2:291 ([ASM]-Clark 2004); and the Daley site (AZ CC:2:235[ASM]-Clark 2004; EAC 1981). It is conceivable that the small occupation exposed at AZ CC:2:289(ASM) (Hall and Clark 2004: 23-41) was also using this canal. A large artifact scatter (AZ CC:2:236[ASM]-Clark 2004) also may have been associated. Other sites (e.g., the "Safeway site"), for which I have been unable to obtain information, were also probably associated.

## The Sunflower Canal

The Sunflower Canal is proposed as the fifth canal on the south side of the Gila River that may have had a prehistoric counterpart. The 1985 *Safford, Ariz.* quadrangle of the USGS 7.5-minute Topographic map series indicates that the Sunflower Canal has its present head or offtake at a well and is surrounded by fields. Our survey found this canal to have an unlined dirt channel about 1.5 m wide at its floor and one meter deep, but could not locate its head at the Gila River. However, an interview with Frank Quinn (Neely, personal communication 1997) and the 1920 map (ASWC 1920) permit the placement of the historical, and possible prehistoric, offtake from the river just southwest of, and across the river from, site AZ CC:2:10(ASM) (see Figures Seventeen and Eighteen). We were able to follow the course of the Sunflower Canal towards Thatcher, but we lacked the time to trace it to its termination.

In 2000, archaeological testing by Desert Archaeology, Inc. discovered prehistoric canal segments and associated habitation sites north of U.S. Highway 70, between the communities of Safford and Thatcher (Clark 2004; Nials et al. 2004). The excavations disclosed segments of two ancient canals (AZ CC:2:296 and 297[ASM]) that generally parallel the Sunflower Canal and are only a short distance to its north (see point "D" on Figures Seventeen and Eighteen). The dimensions of the prehistoric canals (Nials et al. 2004) are not quite as great as the larger canals found at Epley's Ruin, however, they are larger than the Sunflower Canal is today. These prehistoric canals have been radiocarbon dated at A.D. 1-300 and A.D. 900-1400. Early historic canals also found during the excavations indicate a long, continuous history of canal irrigation.

*Sites Probably Associated with the Ancient Predecessor of the Sunflower Canal.* Four or five habitation sites were probably associated with the ancient counterpart of the Sunflower Canal: AZ CC:2:10(ASM); sites AZ CC:2:290(ASM); and AZ CC:2:291 ([ASM]-Clark 2004), and probably the Daley Site (AZ CC:2:235[ASM]-Clark 2004; EAC 1981). While site AZ CC:2:289(ASM) is only listed as an "agricultural activity" site (Clark et al. 2004:Table 1.1), the occupants of its two pit structures (Hall and Clark 2004: 23-41) may well have utilized this canal.

## Prehistoric Canals North of the Gila River

Our survey on the north side of the Gila River has just begun, and therefore is largely incomplete. To date, the Michelena-Tidwell and Graham Canals are the only canals on the north side of the Gila River we believe to have evidence of possible prehistoric counterparts (Figure Seventeen). Our evidence consists of historical information and the proximity of a number of prehistoric sites, and is more circumstantial than that for the canals on the south side of the Gila River.

### The Michelena-Tidwell Canal

The canal presently known as the Tidwell Canal is located on the north side of the Gila River, north of the community of San José (Figure Seventeen-Colvin 1997, 1998; Colvin and Cook 2006:11). This canal was first known as the Michelena Canal, and perhaps also included the Brown, Mejia, and Sanchez Canals in its history. A letter written by R. Louis Michelena (1990) indicates that his uncle's father and grandfather, Manuel and Trinidad Michelena, excavated the Michelena Canal around 1853. Assuming the 1853 date is correct, this makes this the earliest historic canal known for the Safford Basin. Because the Michelena Canal is officially registered as having been constructed in either 1874 (Colvin 1998:25) or 1885 (Michelena 1990), this letter also illustrates the early practice of first excavating a canal, and then officially registering it several years later; a situation also noted regarding the Fourness Canal by Colvin (1998:21). Because of the documented practice of the early Hispanic farmers refurbishing prehistoric canals for use, and the proximity of archaeological sites to this canal, we believe it is likely that this canal may have had a prehistoric counterpart.

Because of a number of refurbishments, that may well have also lengthened the canal, the head or offtake of this canal from the Gila River is difficult to place. The original head of this canal may have been across the river from that of the Old San José-Fourness Canal.

*Sites Probably Associated with the Ancient Predecessor of the Tidwell Canal*. The four habitations sites (from east to west) that may have been associated with the proposed ancient canal are: the Earven Flat Site (AZ CC:2:8[ASM]); AZ CC:2:5 ([ASM]-Ahlstrom 1997; Brown 1973; Tuohy 1960); AZ CC:2:6(ASM); the Buena Vista (Curtis) site (AZ CC:2:3[ASM]-Brown 1973; Fewkes 1898, 1904; Mills and Mills 1978; Tuohy 1960); and AZ CC:2:9(ASM). The Sanchez agricultural sites (Seymour et al. 1997) may also have been associated.

## The Graham Canal

To the west of the Tidwell Canal, the second canal on the north side of the Gila River we believe may have had a prehistoric counterpart is the Graham Canal. Part of the circumstantial evidence is historic information dealing with the town of "Smithville" (now named Pima), and recorded in Mormon Church records in the "St. Joseph Stake History, Pima Ward" and "St. Joseph Stake History, Eden Ward" (Williams 1937:23). According to Williams (1937:23), the document states that the Mormons: ". . . had enlarged fifteen to twenty miles of the old ditches (the old ditches were widened from three to four feet to, in many cases, eight to ten feet and deepened proportionally)." A second historic document (Talley 1970:7) states that "... an old channel ..." was used in the process of realigning and constructing a new diversion for the Graham Canal after the disastrous flood during the winter of 1905-1906. While the "old ditches" and "old channel" mentioned in these documents could refer to earlier historic channels, they could equally well refer to prehistoric canals. There are also a number of prehistoric sites located along the course of the Graham Canal.

*Sites Probably Associated with the Ancient Predecessor of the Graham Canal*. At least four habitation sites that may be associated are (from east to west): AZ CC:2:10(ASM); the Peterson Wash Site (AZ CC:2:31[BLM]-Taylor 1983); the Peck Wash Site (Neely and Doolittle 2004:131-132); and AZ CC:1:3 ([ASM]-Neuzil 2005b). Although primarily focused on agave cultivation, the Safford Grid site (AZ CC:1:2[ASM]-Doolittle and Neely 2004; Neely and Doolittle 2006) also has small habitation components that may have been associated. Off the map (Figure Seventeen) to the west (see Neely 2008a:Figure 1) there are five more habitation sites that may well have been in association, these are:

AZ CC:1:5(ASM); AZ CC:1:6(ASM); the Eden site (AZ CC:1:4[ASM]-Tuohy 1960); AZ CC:1:7(ASM); and the Owens-Colvin site (AZ CC:1:19[ASM]-Neily et al. 1993; Rule 1993).

## Observations

A number of observations have resulted from our study of the canals in the eastern half of the Safford Basin. These observations are pertinent to provide perspective and better understanding of the prehistoric Gila River canals and related phenomena.

1) The re-excavation of ancient canals, and/or the positioning of historic canals over or adjacent to the channels of ancient canals, in the Safford Basin should not be considered unusual or isolated events. A growing body of evidence suggests that the refurbishment and reuse of prehistoric canals was rather commonplace (Doolittle 2000:368-369). Such refurbishments have been documented for the prehistoric canals of the Middle Gila Valley (Haury 1976:122-123; M. Kyle Woodson, personal communication 2006), the Phoenix Basin (Masse 1981; Purcell 2007; Rogers 1976), and throughout the world (e.g., Peru [Gelles 1996]; Sri Lanka [Stanbury 1996]; Sonora, Mexico [Doolittle 1988]; Puebla, Mexico [Neely 2001b, 2005b; Neely and Rincon Mautner 2004, Woodbury and Neely 1972]).

The seven canals postulated to represent prehistoric features in the Safford Basin have apparent prehistoric counterparts in other areas in what is now southern Arizona. For example, the canals found in the Middle Gila Region in the vicinity of the Casa Grande ruin by Crown (1984, 1987) were excavated in similar topographic locations, and in similar relationships with sites and fields, as those occurring in the Safford Basin. Although farther removed, the pattern of prehistoric sites vis-à-vis canals is also generally similar to that in the Phoenix area (e.g., Abbott et al. 2006:Figure 1; Howard 2006; Howard and Huckleberry 1991; Hunt et al. 2005:Figure 3).

2) The early radiocarbon dates recovered from the prehistoric canals of the Safford Basin are not unusual considering the much earlier dates being recovered from agricultural sites and canals in the Tucson Basin (Mabry 2007; Theil and Mabry 2006).

3) In reviewing the locations of the sites and canals being considered in this paper, two interesting things may be seen. First, each of the large major sites is bounded on two sides by water sources (Figure Eighteen) Epley's Ruin by

the Montezuma Canal and the San José Canal; AZ CC:2:2(ASM) by the Union Canal and the Montezuma Canal; the Methodist Church Site by the Union Canal and both the Montezuma Canal and the San José-Highline Canal; the large unrecorded site south of the Highline Canal by that canal and several springs; and possibly the Buena Vista site by the Gila River and the San José Canal. Second, the canals have been engineered to maximize the available irrigable lands by conveying water from their offtakes on the T1a floodplain terrace and channeling it up onto the T1b terrace (Figure Eighteen). Thus, apparently each of the canals was designed to supply water to segments of the terraces that could not be supplied by the other canals, either because the fields lay up-slope or because there was not sufficient water to irrigate all of the field area lying between a higher canal and the river.

4) The San Jose-Highline Canal is extremely well engineered. It follows a course that keeps it at the highest point a prehistoric canal probably could have been excavated in the southeastern part of the Safford Basin. This provides irrigation water to the highest irrigable area to its west and north and, thereby, maximized the use of the southern Gila River floodplain. This engineering design is apparently quite similar to that used in the construction of the canal that supplied water to the area of the Casa Grande Ruin (Crown 1987:Figures 2, 3; also see Doolittle 2000:390-392). The proposed prehistoric San Jose-Highline Canal is, therefore, in keeping with the prehistoric engineering skills demonstrated outside of the Safford Basin.

In addition, considering the choice placement of the San José-Highline Canal at the toe of the T3 terrace, it would not perhaps be surprising to find an earlier similar long canal following the toe of the T1b terrace, which is generally located just north of and follows the present route (from east to west) of the lower San José Road and U.S. Highway 70 (Figure Eighteen). The placement of a canal in the latter location would conform to the upper limits of irrigable floodplain fields proposed by Doolittle (2008). While there is no direct evidence to support this hypothesis, it is a scenario that future researchers should keep in mind.

5) Because we are uncertain of the original dimensions of the prehistoric primary canals at any time during their periods of function and the systemic relationship of the canal segments exposed through excavation, we have demurred from categorizing the excavated examples as being primary, secondary, or tertiary in nature. Some 145-airline kilometers to the

west-northwest of the city of Safford, in the vicinity of Florence, Arizona, are the closest recorded remnants of major prehistoric canals coming from the Gila River outside of the Safford Basin. Crown (1987:150-151; also see Doolittle 2000:390-392) reports the presence of a single large prehistoric primary canal on the north as well as the south bank of the Gila River. The north canal is said to have a discernable length of 11 km and a width of 12 meters, while the south canal was measured at 32 km in length and has a width of four to seven meters. Both of these canals were as long and as wide as many of the larger historic Gila River canals found presently in the Safford Basin. With these comparative data, supported by size comparisons with canals in the Phoenix Basin (Abbott et al. 2006:285; Doolittle 2000:396-408; Howard 2006), the large size of the San José Canal, and its inclusion of what is now termed the Highline Canal, is reasonable for a large Southwestern prehistoric primary canal. Thus, the size of the San José-Highline Canal rather closely approximates Crown's canals for at least one stage in their functioning existence. In this light, the tentative categorizations of "secondary" and "tertiary" (Huckleberry [2005] and Nials et al. [2004]) for the excavated prehistoric canal segments may well be correct. However, the large canals in the Casa Grande-Florence area may only represent the later occupations (ca. A.D. 1150-1450), which could well affect the interpretations of the Safford Basin prehistoric canals. For example, the relatively large size of the Epley's Ruin Feature #6 canal (Huckleberry 2005) and its early date (190 B.C.-A.D. 10) may indicate one of two things, pending further investigations. It could represent an early primary canal, rather than a secondary canal as it has been attributed. Or, Feature #6 was, in fact, a segment of a secondary canal; which indicates that at least one of the early canals of the Safford Basin was quite large. Here we are faced with two interrelated dilemmas: lack of broad excavation exposure and a single very early radiocarbon date, neither of which lends itself to very secure interpretations. Very much as Doolittle (2000:408) has noted for the Phoenix Basin, and we have observed in the realignments of historic canal channels, the canal systems and the irrigated landscapes of the Safford Basin should be considered as dynamic and ever changing (Huckleberry 2005). In certain times and situations smaller canal systems would have been suitable, while in others larger systems would have been more appropriate and necessary. An interesting comparison may be made between discoveries in the Safford Basin and Hunt's (1994) findings in the Cuicatlán area of highland central Mexico. Hunt's examination of the Río Grande provides striking similarities with the Gila River. Yet, there is no evidence that the Río Grande of the Cuicatlan Valley had been used for canal irrigation in prehistoric times. In a very thorough examination of the data, Hunt (1994) concludes that the probable explanation for this lack of canal irrigation was the fact that the prehistoric agriculturalists had no need for such technology. However, Hunt (personal communication 2006) observed that the lack of such evidence is not evidence that the river was not used. Obviously, the prehistoric agriculturalists of the Safford area deemed canal irrigation from the Gila River very necessary in spite of the constant threat of crop destruction and canal washouts by climatic vagaries affecting the river (Neely 2004).

6) We have recorded a broad band-like area of springs with an eastern edge beginning about five kilometers west of the juncture of the San José Canal with the Highline Canal. These springs occur in a general south-southwest – north-northeast alignment from the foothills of the Pinaleño Mountains to the scarp of the T3 terrace above the river floodplain. This alignment of springs is approximately five kilometers in width, and probably indicates the presence of a geologic fault. The overflow from these springs drains northward into the Highline Canal. The archaeological sites located in this area were most likely situated to take advantage of these springs. The northernmost of these sites (e.g., AZ CC:2:69[ASM] and the unrecorded site now covered by a subdivision of Safford) may also have been associated with the Highline Canal and the fields to the north that it serviced. As the general topography slopes from the Pinaleño Mountains towards the Gila River, runoff from rainfall and snowfall would have also augmented the canal flow.

7) Unlike what has been found with the dry-farming (Doolittle and Neely 2004; Neely 2001a, 2008a; Neely and Doolittle 2006) and foothill (bajada) irrigation (Neely 2001a, 2005a, 2008a, b) systems of the Safford Basin, the fields associated with the Gila River canals remain poorly known. This is due to the great modifications of the Gila River floodplain landscape by intensive historic cultivation (which includes ongoing mechanical land-leveling) and the urban expansion of Safford and other communities paralleling the Gila River since the late 1800s.

Our findings correspond with and are reinforced by Doolittle's (2008) reconstruction of the potential agricultural area of the Gila River floodplain. Additionally, the possibility that the San José-Highline Canal may have serviced sites and fields at higher elevations (i.e., on the T1b level of the Gila River floodplain [Figure Eighteen-Huckleberry 2005:Figure 3.2]) would, perhaps for the later periods of agricultural activity, have substantially increased Doolittle's (2008) area estimates of lands that were available for irrigation by Gila River-based canals. Doolittle has estimated that as much as 7,000 hectares (17,290 acres) of the floodplain were available for irrigated cultivation along both sides of the Gila River in prehistoric times, but as yet we have found nothing visible on the ground surface to indicate their presence. Nor did the limited excavations of Botsford and Kinkade (1993), Clark (2004, 2006), and Lascaux and Huckleberry (2006-also see Huckleberry 2005) find evidence of the floodplain fields. In sum, we have no information as to their nature, shape or size, because the entire Gila River floodplain is currently being used as agricultural fields, or is beneath urban development. However, there are four sources to which we may refer to obtain some very limited ideas as to the prehistoric fields associated with the Gila River canals.

First, we may glean some idea as to single field shapes and sizes from the well-defined irrigated fields found in Lefthand Canyon (Neely 2005a, 2008a). The apparent fallacy in using the foothill (bajada) systems as a model is the limited water supply and smaller canals present. This would suggest that only the largest fields in the foothill systems possibly approached the sizes of the Gila River floodplain fields. In Lefthand Canyon the largest irrigated fields were found to be rectangular in shape and range from about 6,000 $m^2$ to 14,300 $m^2$ in area (Neely 2005a, 2008a).

Second, as in other regions (e.g., the Tehuacan Valley of Puebla, Mexico [Neely 2005b] and the Near East [Wilkinson 2003]), it is reasonable to presume the ancient cultivators of the Safford Basin Gila River floodplain were required to modify, develop, and manage the landscape to effectively utilize canal irrigation. In many regions these prehistoric modifications of the landscape to accommodate canal irrigated agriculture were maintained and continued into the historic period. If this is extrapolated to the Safford Basin, then we may assume that the historic Gila River floodplain fields are in some respects similar to those used in prehistoric times. While this may be true to some degree, we suspect that it is so only in the grossest terms (see Doolittle 2006).

Third, Bandelier presents a color drawing of a canal segment from a location to the west of Safford near Fort Thomas (Burrus 1969:83, 196). As may be seen in Figure Twenty Seven, Bandelier has provided information as to the dimensions of this canal segment and the distances between branching canals. The distance between branching canals presumably provides one dimension of a single field. The fact that the branching canals leave the main canal at nearly a right angle suggests that at least some fields were quadralinear in shape; information that is verified by findings in Lefthand Canyon (Neely 2005a). Furthermore, the representation of some of the smaller canals with nearly right-angle bends in their channels shortly after branching from the larger canal suggests a manner by which water flow was probably slowed in the channels. With the data at hand, the Bandelier drawing indicates that the fields were probably rectangular in shape, but the great variation in the field dimensions is perplexing; the smaller dimension overlaps with one dimension of some of the Lefthand Canyon fields, but the larger dimension greatly exceeds the maximum dimension of the largest of the Lefthand Canyon irrigated fields. However, this illustration also generates other questions. For example, is the larger canal shown a primary or a secondary (distribution) canal? Did this canal carry waters from the Gila River, or is it part of a spring-fed foothill (bajada) system? While this canal may have been a segment of the same canal reported by Bandelier (1892:410; also see Hough 1907:38 and Neely 1998a) as: "About eight miles east of Fort Thomas . . . ", and descending " . . . from the base of the foot-hills of Mount Graham . . . ", its size exceeds that of the foothill canals. The orientation Bandelier gives does not correspond well with canals coming from either the foothills of the Pinaleño Mountains or from the Gila River.

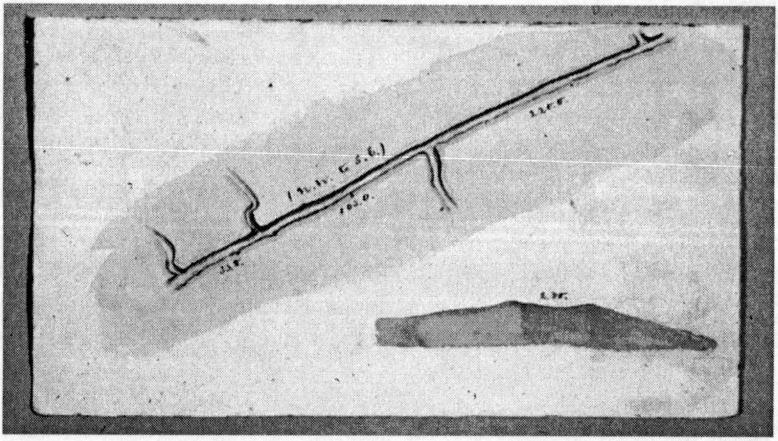

Figure 27: Plan and profile drawings of a canal segment with branching smaller canals found by Adolph Bandelier between Safford and Fort Thomas in the Safford Basin. From Folio #103 of the watercolor drawings by Adolph F. Bandelier in the Vatican Library (Burrus 1969:196).

The final, and perhaps currently the best, resource we may draw from to obtain information that would possibly assist us in obtaining shape and size information about the Safford Basin floodplain fields is from outside the basin. This resource, although it too is extremely sparse, pertains to the fields paralleling the canal systems in the Phoenix Basin (Howard 2006; Howard, personal communication 2007). Howard has determined that some of the Phoenix Basin canal irrigated fields in Canal System #1 ranged from around 1.8 to 5.0 acres (7,285 to 20,235 $m^2$), a good match for the available data on early Akimel O'odham (Pima) field size (Russell 1908:87-88). It is interesting to note that this range of field areas is quite similar to that noted above for the Safford Basin's largest foothill irrigated fields. Howard also determined that in some areas of the Phoenix Basin there were from 15 to 40 fields serviced by the secondary (distribution) canals, again a rather good match for many of the ethnohistoric areas under irrigation discussed in the literature.

An additional pertinent observation is that: before the excavation of these prehistoric canals and the cultivation of the fields, the ancient agriculturalists probably had to expend a good deal of time and effort in removing the dense mesquite *bosques* that characterized the floodplain (Doolittle 2006:182-184; Ramenofsky 1984:50-57). While this clearing provided the land necessary for agricultural intensification, it also destabilized the river regime and increased the probability of geomorphic change that undoubtedly resulted in "floodplain stripping" and the washing out of canal headworks (Doolittle 2006:182-184). On the other hand, the cleared areas would have received alluvium carried by the waters of the canals. This would have added rich, rejuvenating sediments to the cultivated areas as well as replenishing the to deposits lost to "floodplain stripping", and possibly acted as a stabilizing factor (Doolittle 2006:190-191).

8) The proximity of the proposed prehistoric canals to prehistoric occupation sites (Figures Seventeen and Eighteen), and the fact that some of these canals actually passed through sites, strongly indicates that the canals served as domestic water sources as well as for agricultural irrigation. This is particularly evident for the large Buena Vista (Curtis) site (AZ CC:2:3[ASM]), Epley's Ruin (AZ CC:2:64[ASM]), AZ CC:2:2(ASM), and the Methodist Church site (AZ CC:2:15[ASM]). Although evidence of its existence is now difficult to find, the Methodist Church Site, in the center of the present city of Safford and about two kilometers south of the Gila River, was evidently one of the largest sites in the Safford Basin. This site is

conservatively estimated as once covering an area of about 735,000 $m^2$ (73.5 hectares). The present course of the Union Canal apparently passes through, or immediately north of, this site. Fewkes (1898:610-616, 1904:170-175) also mentions the presence of an abundance of prehistoric canals and a multitude of mounds between the Buena Vista site and Solomon. He relates that his 1897 visit to the Safford basin found most of these sites already highly disturbed by agriculturalists. He also notes that interviews with residents indicated that these sites were still well preserved just 20 years earlier (i.e., 1877). The practice of using canals to supply domestic water is also indicated by the relative locations of sites and canals in the Middle Gila (Crown 1987:Figure 2; Haury 1976) and Phoenix (Abbott et al. 2006:Figure 1; Hunt et al. 2005:Figure 3) areas.

9) The placement of settlements in proximity to fields and water management infrastructure in all of the environmental sub-zones within the basin (Neely 2008a) facilitated the planting, tending, and harvesting of crops and maintenance of water management systems. In addition, it placed habitation in varied natural and human-developed ecotones (Clements 1904; Laurance et al. 2001; Odum 1983), making available greater numbers and varieties of wild plants and animals to supplement the diet and for other uses. For example, the prehistoric inhabitants of the Safford Basin very probably utilized the dense mesquite forests on the floodplain of the Gila River for wood and mesquite pods (ground for food).

Gila River canals, as well as canals having other water sources, very likely generated a human-made riparian microenvironment along their courses. Such riparian zones would have provided a ready supply of edible and useful plants, and would have been a haven for animals. While historic construction and especially cultivation have negatively affected the presence of this microenvironmental zone along most portions of the Safford canals, there are sections of the canal routes that are characterized by heavy vegetation with multiple species of arboreal and non-arboreal plants present. This phenomenon is clearly evident with the lower portion of the San José-Highline Canal where it courses along the toe of the high T3 terrace (Figure Eighteen). Although not mentioned previously (i.e., Neely 2005a), the smaller prehistoric canals distinguishing Lefthand and Marijilda Canyons are also set apart by a relict riparian microenvironment paralleling their courses. When Neely mapped the canals in Lefthand Canyon, the alignments of mesquite trees and vegetation were used as a guide in tracing the poorly defined segments of the canals. This riparian phenomenon has also been observed elsewhere in Arizona (e.g., Purcell

2007; David Purcell, personal communication 2007) and along the "fossilized" canals in the Tehuacan Valley of southern Puebla, Mexico (Caran and Neely 2006; Neely 2001b, 2005b).

10) It appears that the terrace on which the Buena Vista site is located may have been partially altered to convey water from rainfall and snowmelt outward from the site over its perimeter to the canal and fields below. Such site and landscape modifications to harvest water have been recognized for archaeological sites in the Maya region (Scarborough 1993, 1994; Scarborough and Gallopin 1991; Scarborough et al. 1995) and in southern highland Mexico (Neely 1972, 2005b), and may have been more widely practiced in the American Southwest than has been perceived.

## Temporal and Ethnic Placement of the Safford Basin Gila River Canals

Our search for written primary historical documents relating to the very early Gila River fields and canals was only moderately successful. Our earliest written resources were the reports of Babbs (1901), Fewkes (1898, 1904), and Southworth (1919). The files of the Gila River Irrigation District began in 1920, when the agency was founded. In a recent telephone conversation with Verna Rae Colvin (personal communication 2006), the recognized expert on the historic canals of the Safford Basin, we learned that she had found no earlier primary written sources. Her information evidently largely stems from interviews with elderly residents of the Safford area, from interviews with and articles by Ryder Ridgway, a long time Safford resident and historian, and from the files of the Gila River Irrigation District. It was primarily through interviews with her grandfather that Elizabeth Ramenofsky obtained the majority of her historical information regarding the fields and canals in the Safford area (Ann F. Ramenofsky, personal communication 2006), although she too evidently referred to Ryder Ridgway for information (Elizabeth Ramenofsky Collection, MSS #117, Arizona Historical Foundation).

The temporal and ethnic attributions for the historical Safford Basin Gila River canals have been documented. Colvin (1998) states that, in 1872, the Hispanic Mejia, Monte, and Montoya families excavated the Montezuma Canal, the first officially registered historic canal in the Safford area. In 1874, Francisco Monte and other Hispanic families registered the San José Canal. Eleven other canals were excavated between 1874 and 1895 (Colvin

1998:21-28), most by men with non-Hispanic surnames, many of whom were members of the Church of Jesus Christ of Latter Day Saints (the Mormon Church). As noted above, the historic information may not be as accurate as it is assumed it to be. We suspect that sometimes there is an interchangeable use of the words "excavated" and "registered." As noted for the Michelena-Tidwell and Fourness Canals discussed above, the unexplained gaps in time between when canals were "excavated" and then later officially "registered" is misleading. Unfortunately, most of the possible incongruities occur in relation to the earliest historic information, apparently due to the lack of clearly written documentation.

Other historic information indicates that at least some of the Safford Basin canals were originally excavated much earlier than their officially recorded historic dates. Colvin claims prehistoric origin for some of the Safford Basin canals. She relates that both the Montezuma and San José Canals had been historically excavated: " . . . following the canal system that had been built by the Hohokams" (Colvin 1998:6), and that they: " . . . followed an old irrigation system that was used by pre-historic Indians" (Colvin 1998:20-21). The prehistoric origin of these canals is also supported by Fewkes (1898:613, 1904:178) and Ramenofsky (1984:42); who writes that in 1876 her Grandfather, Isadore E. Solomon, one of the first non-Hispanics to enter the Safford Basin and the founder of the town of Solomonville, found Hispanic farmers using "ancient" canals that they had re-excavated to irrigate their crops. Having provided credible archaeological and historical probability for the prehistoric origins of the Gila River canals discussed herein, we now consider by whom and when these prehistoric canals were excavated.

A review of the more recent literature (e.g., Botsford and Kinkade 1993; Clark 2004, 2006; Crary 1997; Doolittle and Neely 2004; Huckleberry 2005; Neely 2001a, 2005a, 2008a; Lascaux and Huckleberry 2006; Lascaux and Montgomery 2005; Neuzil 2005b, 2006a, 2006b; Purcell 2006b; Rinker 1998; Seymour et al. 1997; Woodson 1995, 1999, 2006), involving both recent survey and excavation, quickly reveals that ceramics that have been attributed to five prehistoric "cultural" or "ethnic" groups have been found in the Safford Basin. While artifacts from earlier inhabitants are present, the San Simon Mogollon (Sayles 1945) was the first group to produce culturally distinctive ceramics in the Safford Basin (but see Di Peso 1956, 1979; Whittlesey 1995). However, four other Southwestern prehistoric "cultures" or "ethnic groups" are later represented by ceramics in the Safford Basin. In the apparent order of their appearance, ceramics attributed to the Hohokam, Mimbres Mogollon, Anasazi, and Salado are represented. In light of the little we know about Safford Basin archaeology, and the large number of sites that have substantial quantities of ceramic types attributed to several of the "cultural" or "ethnic" groups noted above, we have wondered if what we are seeing is a reflection of heavy ceramic trade rather than the incursion of several multiethnic groups. This question appears to have been resolved in the article by Purcell and Clark (this volume), who report ethnically diverse contemporaneous occupations at two sites in the Safford Basin. Considering the findings of Purcell and Clark, many details must be investigated: what were their numbers, what were their roles in the socio-political schema of the basin, as well as what economic and socio-political relationships they had with the San Simon Mogollon and among themselves? Were they traders that subsequently formed sedentary enclaves (e.g., Clark 2001; Neely 1974; Neuzil 2005b), or did they come for other reasons? There is recent evidence from the Goat Hill site (AZ CC:1:28[ASM]-Woodson 1995, 1999, 2006), and possibly other sites in the Safford Basin (Neuzil 2005b, 2006a, 2006b), that Anasazi groups from northeastern Arizona, evidently passing through the Point of Pines region of east-central Arizona, came to the Safford Basin and established residency during the period of ca. A.D. 1275-1325. Climatic change and related economic and socio-political problems are suggested as the most likely reasons for this migration. It is not unreasonable to consider these as among the causes for other prehistoric groups to visit and perhaps settle in the Safford Basin. Now that such incursions with residency have been documented, the Safford Basin presents a unique opportunity for a longitudinal study of several prehistoric "ethnic" groups with overlapping periods of coeval residence.

To place these occupations of the Safford Basin in a temporal framework, the cited literature notes that the same ceramic types also serve as chronological markers indicating an occupation time span beginning perhaps as early as ca. A.D. 200 and lasting until about 1450. Even earlier and more conclusive dates for the occupation of the Safford Basin, and for the early construction and use of sophisticated water management and irrigation technology, come from radiocarbon samples collected from Gila River canal contexts. The excavations of Desert Archaeology (Clark 2004, 2006; Nials et al. 2004) disclosed two canals dating between A.D. 1 and A.D. 300. The Epley's Ruin excavations (Huckleberry 2005; Lascaux and Huckleberry 2006; Lascaux and Montgomery 2005) revealed an even earlier canal, radiocarbon dated at between 190 B.C. and A.D. 10.

## Summary and Conclusions

The Safford Basin has not received the archaeological attention that has been given other areas of the American Southwest. However, over about the last dozen years several projects have been conducted, and this largely ignored area is finally being recognized for its importance and potential. Most of this work has been survey, and excavation is sorely needed to verify the models generated and to provide the details that are not available through survey. Due to the foregoing, most of the following conclusions must be regarded as only initial considerations.

The archaeological and historical evidence strongly suggests that several of the historical canals taking water from the Gila River were either prehistoric canal refurbishments and/or closely follow the routes of prehistoric canals. Similar evidence from other regions of the American Southwest and elsewhere in the world support this practice, and therefore increase the feasibility of this having occurred in the Safford Basin.

Although the recent excavations in the basin (e.g., Clark 2004; Lascaux and Montgomery 2005; Purcell 2006b; Rinker 1998, and Woodson 1999) have provided some information as to the cultural continuity and change in the basin, we are still woefully ignorant. Ideally, large-scale excavations in floodplain contexts are needed to augment the existing small sample of sites and canals. Additional radiocarbon dates are also needed to more securely date features as well as to clarify their contexts, and to discover their sequence of construction and lengths of operation. During this process it seems likely that we would also learn a great deal more about the fields these canals serviced.

The discovery of Gila River-supplied canals with very early radiocarbon dates during the excavations directed by Clark (2004, 2006) and Lascaux and Huckleberry (2006; Huckleberry 2005), especially those at Epley's Ruin (Lascaux and Huckleberry 2006; Lascaux and Montgomery 2005) which were in association with Mogollon household refuse, ceramics, architecture, and other features, suggest that the San Simon Mogollon planned, excavated, and used the first Gila River canals in the Safford Basin. The span of prehistoric canal use evidently extends from perhaps as early as 190 B.C., but at least from ca. A.D. 10 to A.D. 1450. However, the evidence is very limited and sorely needs augmentation.

That the Mogollon may have been the initial canal builders in the basin may be considered unusual due to the relative sparseness of canal construction in the Mogollon heartland of west-central New Mexico and east-central Arizona, and the relatively small scale of those that have been recorded (e.g., Damp et al. 2002; Neely 1995). Be that as it may, the sparseness and small scale may well be due to as yet unperceived cultural factors, as well as the environmental conditions and the near absence of major streams in those regions. It seems unlikely that we will ever be able to determine if the early canals in the Safford Basin were a result of independent invention or emulation. It is entirely possible that the early canals in the Tucson or Phoenix Basins may have provided the prototypes for the Safford Basin canals. However, it is perhaps more appropriate and productive to attribute canal technology development to environmental conditions (i.e., availabilities and limitations) as well as local group necessities rather than resulting from direct contact with or influence from any specific "cultural" or "ethnic" group.

While the San Simon Mogollon appear to have been the originators of Gila River canal engineering in the Safford Basin, subsequent canal network modifications and expansions were likely influenced by later inhabitants; be they the descendants of the San Simon Mogollon or other groups that subsequently occupied the basin. It is feasible that the Hohokam may have had a role in canal enlargement and network expansion, but, if they did, how much of a role they had and how much other occupants of the basin contributed is unknown at this time. As suggested by the historic efforts to refurbish the prehistoric canals and subsequently enlarge the canals and expand the canal network, we think it appropriate to model the continuing prehistoric enlargements and expansions as an incremental process (Doolittle 1984).

Our perspective on canal engineering, as well as other aspects of water management, should be viewed in the same manner as Doolittle and Mabry (2006) have modeled the development of agriculture in the American Southwest. In the following quote, the replacement of the words "maize" with "water management technology" and "evolutionary" with "developmental", would modify it to appropriately fit our thoughts on water management development, including floodplain irrigation.

> Rather than thinking in simple, specific, single-process, single-event terms, our thinking about the early history of maize in the greater Southwest

> should be framed in the context of complex, diverse, and evolutionary processes over an extended period of time. (Doolittle and Mabry 2006:118).

Survey has shown that substantial numbers of fields and other agricultural infrastructure are present in the three major environmental zones of the Safford Basin (Neely 2004, 2008a). It has been proposed that such diversification of agriculture most likely reflects a strategy to reduce the risks resulting from the highly variable and unpredictable water/moisture resources (Neely 2004). Survey also suggests that the scale and number of field areas and water management infrastructure increase through time. As with the use of multiple environmental zones, the increasing construction of labor-intensive water management systems probably represents an adaptive risk-reducing strategy. Complimenting and augmenting these strategies was the broadly distributed placement of habitation sites, which facilitated access to agricultural fields and water management systems as well as to an increased variety of naturally occurring plants and animals. It is conceivable that the foregoing strategies may also have been instituted to ameliorate socio-political, and perhaps ceremonial/religious, conflicts arising from an increasing population of local as well as ethnically, and probably linguistically, diverse immigrant peoples.

Socio-political organization must also be considered in the light of the increasing human modifications of the landscape, for it is only through the organized efforts of groups that large-scale projects could be accomplished to increase the availability and predictability of moisture. Although the prehistoric agricultural strategies of the Safford Basin apparently involved relatively large areas and complex infrastructure, they have been modeled as products of a group, or groups, with a kin-based organization that was strengthened by cross-cutting ceremonial affiliations (Neely 1997b, 2005a, 2008a; Neely and Doolittle 2004). Aspects of that organizational model have received additional support from a recent publication considering the plausible ethnographic analogies for the social organization of Hohokam canal irrigation (Hunt et al. 2005), as well as studies of Hohokam community organization and exchange in the Phoenix Basin (Abbott 2000; Abbott et al. 2006). Now that some aspects of the prehistoric agricultural strategies of the Safford Basin have been investigated and are at least partially known, we need to seek information to understand the interrelationships of the agricultural system with the economic, socio-political, and ceremonial systems.

Some of the many questions that arise from our study, that may only be answered through additional work, are: (1) How did the Gila River canals and fields interrelate with the other prehistoric subsistence and agricultural strategies of the Safford Basin? (2) What was the scale of these canals, were they enlarged incrementally through time, and were there several prehistoric canals in operation contemporaneously? (3) When, and by whom, was each of these canals engineered and excavated? (4) Was each of the major canal systems engineered, excavated, and maintained by a specific socio-political and/or ceremonial/religious entity?

The study of cultural continuity, change, and process involve the recovery and utilization of good qualifiable and quantifiable data. Although the present corpus is presently limited, we have seen that such data are available from the Safford Basin. This paper constitutes an initial attempt to study one aspect of agricultural intensification, as achieved through Gila River-based canal irrigation. Although it is only one aspect in the study of agriculture as process, such studies are important in understanding the development of, and regional variations in, prehistoric subsistence strategies in the Safford Basin and throughout the American Southwest.

# CHAPTER FOUR

# SOJOURNS IN THE FOOTHILLS: THE ARCHAEOLOGY OF THE GILA MOUNTAINS FOOTHILLS NORTH OF THE SAFFORD VALLEY

JOSEPH A. EZZO

## Introduction

In 2005, SWCA Environmental Consultants undertook investigations at 42 pre-contact Native American sites within the Dos Pobres/San Juan mining district immediately north of the Safford Valley (Figure Twenty Eight; Ezzo 2006). The sites represent the material residue of ancient activity within the project area and manifest on the landscape in a variety of forms, including scatters of artifacts, features such as rock piles, and petroglyph panels. These cultural manifestations transform the natural landscape and imbue it with human significance that encompasses traditions and specific world-views.

The environment and landscape of the Dos Pobres/San Juan project area is challenging, to say the least, with its lack of a predictable water supply, lack of arable land, rugged relief, and relatively scant wild food resources. There is no evidence, for example, of intensive use for activities such as farming, so prevalent in the form of the "waffle gardens" or gridded fields found at AZ CC:2:127(ASM), several kilometers to the east at the edge of the Gila River floodplain (Ahlstrom 1997), or large villages and irrigation canal systems that are prevalent along the Gila River. The high foothills of the Gila Mountains were not a place where people could go to make a living, but rather to undertake specialized and limited types of activity. Therefore, the landscape can be perceived as one used for sojourns, rather than for long-term occupation. Yet within its ruggedness and its limited potential for food resources, it provided a place for ancient people to create a visual representation of their cosmology and their identity, thus marking the territory as their own.

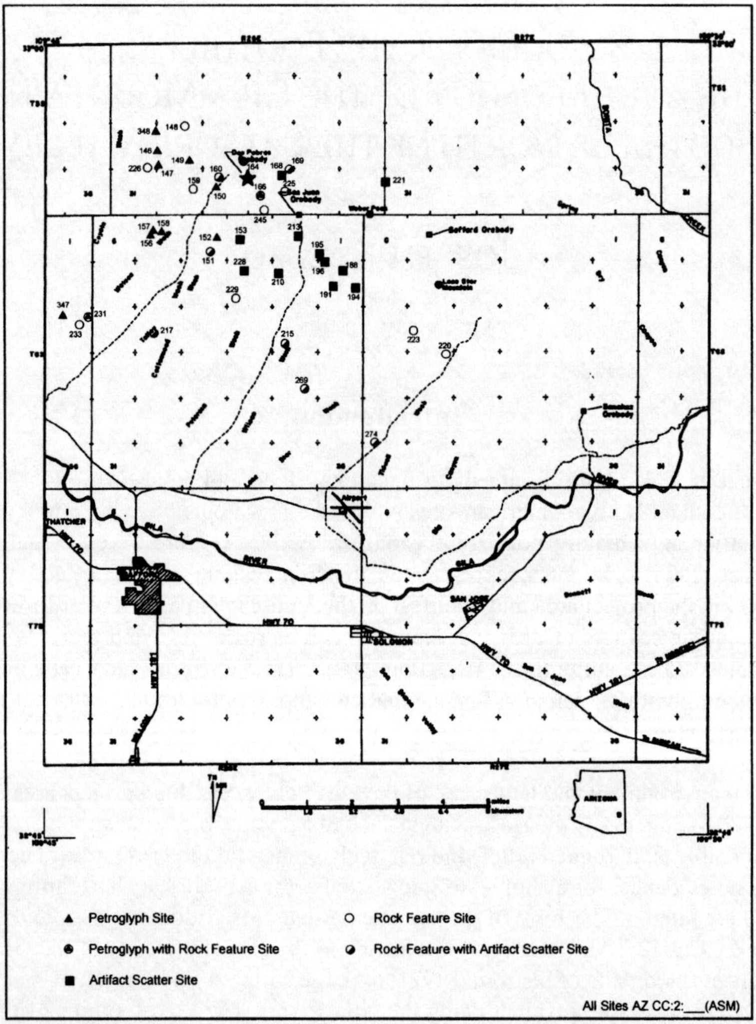


Figure 28: Map of project area showing Native American sites.

As Table Five indicates, the Native American sites in the project area tend to be small and do not exhibit great diversity in terms of features or material culture. Sites consist either of 1) petroglyph panels; 2) rock features; 3) surface artifact scatters; or 4) a combination of petroglyphs or rock features with surface artifact scatters. Site dimensions vary considerably, but there is a certain arbitrary component to defining individual sites. Rock features tend to be scattered about the landscape, often without artifactual association, and determining whether or not features should be grouped as a single site or multiple sites is certainly not an exact science. Consequently, it may be more useful not to analyze the material culture of the project area by discrete "sites," but rather as a relationship between the cultural resources and the landscape. This, however, invites addressing the thorny issue of temporal control, which is tenuous at best. Another approach is to place the sites in a regional context, as was done by Ahlstrom (1997) for the Sanchez Copper project, located less than 5 km to the east of the Dos Pobres/San Juan project area. These approaches will be used here, with particular emphasis on the rock art of the Dos Pobres/San Juan project area.

**Table Five. Native American Site Characteristics**

| SITE NUMBER AZ CC:2:_(ASM) | FEATURES | ARTIFACTS |
|---|---|---|
| 146 | Petroglyph panel | - |
| 147 | Petroglyph panel | - |
| 148 | Possible windbreak | - |
| 149 | Petroglyph panel | - |
| 150 | Petroglyph panel | - |
| 151 | 1 rock pile | Flaked stone |
| 152 | Petroglyph panel | - |
| 153 | - | Flaked stone, Ceramic |
| 156 | Petroglyph panel | - |
| 157 | Petroglyph panel | - |
| 158 | Petroglyph panel | - |
| 159 | 5 rock piles | - |
| 160 | Possible structure, possible hearth, rock feature | Flaked stone, ground stone |
| 164 | Possible roasting pit | Flaked stone, ground stone |
| 166 | Petroglyph panel, rock feature | - |
| 168 | - | Flaked stone, ground stone |
| 169 | Possible roasting pit, 4 rock features | Flaked stone, ground stone, ceramics |
| 188 | - | Flaked stone, ceramics |

| SITE NUMBER AZ CC:2:_(ASM) | FEATURES | ARTIFACTS |
|---|---|---|
| 191 | - | Flaked stone, ground stone |
| 194 | - | Flaked stone |
| 195 | - | Flaked stone, ground stone |
| 196 | - | Flaked stone, ceramics |
| 210 | - | Flaked stone |
| 213 | - | Flaked stone |
| 215 | 2 rock features | Flaked stone, ceramics |
| 217 | 1 rock feature | Flaked stone |
| 220 | 2 rock features | 1 projectile point |
| 221 | - | Flaked stone |
| 223 | 3 Rock features | - |
| 225 | 2 rock features | Ground stone |
| 226 | 1 rock feature | - |
| 228 | 1 rock feature | Flaked stone, ceramics |
| 229 | 2 rock features | - |
| 231 | 4 rock features, petroglyph panel | - |
| 233 | 2 rock features | - |
| 245 | 1 rock feature | - |
| 269 | 12 rock features | Flaked stone |
| 274 | 2 rock features | Flaked stone, ground stone |
| 347 | Petroglyph panel | - |
| 348 | Petroglyph panel | - |

## Thinking in Rock Art

The study of rock art in archaeology, once thought to lie at or near the "lunatic fringe" (Hays-Gilpin 2004), has been increasingly undertaken within the theoretical models that characterize the study of other media within archaeology, such as pottery, textiles, and architecture. A landscapes approach to rock art has been one of the means by which greater cultural context has been achieved. Hays-Gilpin (2004:147) discusses the landscape contexts for rock art, stating that such contributing elements are every bit as important as the artistic images themselves. She lists five such contexts: geomorphological attributes; viewshed; passage of light and shadow; acoustical properties; and plants that grow near the rock art. She also stresses the important territorial information coded in rock art, noting that ancient mobile farmers of Western North America often used rock art to mark prime farmlands (Hays-Gilpin 2004:148).

Many researchers who have studied rock art have described the range of possible functions. According to Wallace and Homlund (1986:112), "[t]he function of rock art, as with any art style, relates to various levels of cultural or social setting." Slifer (2000:18) refers to rock art as an "artifact of ideas" and a "tangible manifestation of beliefs and visions." He elaborates upon this: "[t]here is no doubt that rock art contains messages of many kinds, including ownership, warnings, signatures, demarcations, exhortations, and commemorations as well as stories and myths" (Slifer 2000:18). Utilizing information supplied by Native American informants, Zedeño et al. (1999:xiii) concur:

> . . . [t]he relationship between rock art and the surrounding landforms and the natural and cultural resources give clues as to why people were there, what other resources attracted them to a particular place, and the use intensity of a specific location.

Stone (1991:85) offers a reminder regarding the geographical context of rock art sites:

> In order to assess the function of petroglyph sites within broader social and settlement systems, one should examine the geographic contexts of sites in terms of their access and relationships to natural resources, travel routes, and other sites.

A particular thorny issue when dealing with rock art is the ability of researchers to link rock art panels temporally with archaeological features that are nearby. Wallace and Homlund (1986), in their painstakingly detailed analysis of the petroglyphs of the Picacho Mountains of south-central Arizona, repeatedly emphasize the limitation of interpreting various panels with the archaeology of the area because of the difficulty in establishing temporal control and congruence. Others, such as Schaafsma (1980), Bostwick (2002), and Cole (1990), are equally concerned about controlling the temporal dimension. Ultimately, Wallace and Holmlund (1986:175–176) acknowledge that this is virtually impossible in many cases.

Despite such a powerful caveat, rock art is still extremely valuable as a subject for analysis. It possesses potential meaning at multiple levels of community organization and social life: it may influence aspects of social, religious, ritual, economic, and settlement system, and as such was almost certainly a powerful symbol of identity and cultural affiliation for those who utilized this medium. It imbues the landscape with layers of cultural meaning, often very specific. For example, rock art may be used to locate areas of valuable resources, such as a spring or quarry (i.e., Doolittle 2000; Zedeño et al. 1999). Rock art may also have calendrical or astronomical significance (i.e., White 1994). As previously mentioned, it may have territorial significance, but at the same time carry information regarding cosmology or myths (i.e., Ezzo and Altschul 1993a:41–49; Ezzo and Altschul 1993b:117). As such, information is conveyed on multiple levels, as cosmology and territory became interwoven on both the cultural and physical landscape.

The rock art found within the Dos Pobres/San Juan mine project area include a variety of designs and motifs, including abstract and geometric motifs, anthropomorphic figures, meandering lines, representational groups, and designs that are either unclear or difficult to define. The various motifs are described in turn.

## Abstract and Geometric Motifs

The petroglyphs of AZ CC:2:231(ASM) reside on the northwest (Panel A) and southeast (Panel B) faces of a very small boulder. The spiral on Panel A is a very typical motif that is seen throughout the Greater Southwest and parts of the Great Basin, and have been attributed culturally to the Hohokam, Fremont, Yuman, and various Pueblo cultures (see, for example, Doolittle 2000; Ezzo and Altschul 1993a, 1993b; Schaafsma 1980; Wallace and Holmlund 1986). Schaafsma (1980:83–88) includes this motif as part of the Hohokam Gila style. The Panel B motifs are somewhat indistinct and abstract, but their general linear configuration is also typical of the Southwest (Bostwick 2002; Schaafsma 1980; Wallace and Holmlund 1986).

The petroglyphs of AZ CC:2:348(ASM) are varied and are among the most complex set of motifs in the Dos Pobres/San Juan project area. Three boulders exhibit rock art, including Feature 1, a boulder with motifs on its east (Panel A; Figure Twenty Nine), top (Panel B), and west (Panel C) faces. All three faces are dominated by zigzag or wavy lines. This motif appears in multiples forms on all three panels, with the most intensive expression being on Panel B. Again, these are fairly common motifs in the Southwest, and have been attributed, in the southern Southwest, to having Hohokam affinities (Ezzo 2005; Schaafsma 1980; Wallace Holmlund 1986). Another distinct motif is the grid or rows of connected squares (which is evident on

Panel B), and a variant, consisting of connected squares with a dot in the center, as found in Feature 2.

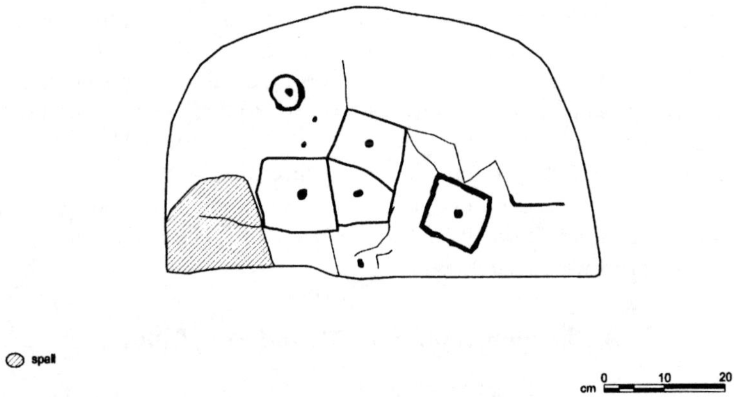


Figure 29: Feature 2, Panel A, AZ CC:2:348(ASM). Note: in Figures Twenty Nine through Thirty Three, differences in shading reflects the depth and clarity of rock art motifs (black indicates deeper and clearer, gray indicates less distinct motifs).

Both sites exhibit geometric and abstract forms, as well as a number of indistinct motifs that cannot readily be identified. Neither contain representational forms, such as anthropomorphic or zoomorphic figures. The most clearly interpretable motif is the spiral at AZ CC:2:231(ASM), which has been identified in conjunction with cosmological significance. It may represent the universe in relation to creation myths, and as a path leading from this life to the next (Hays-Gilpin 2004, Schaafsma 1980). Therefore, there is likely ritual and/or cosmological significance associated with these two sites, and, by extension, the general area around the sites.

The zigzag or wavy lines may likewise represent paths of various sorts, or possibly geographic markers. It is also possible, given the vistas from these two sites, that they represent local mountain ranges. The grid or box pattern may have to do with representing a particular place or specific resource; perhaps. this element marks the actual location or directions to needed resources. Hayden (1972) has interpreted the co-occurrence of zigzag lines, spirals or circles, and stick anthropomorphic figures in rock art along trails of the Papagueria as shrines marking routes of Hohokam shell expeditions to the Gulf of California. These elements do not co-occur here, nor were any shell artifacts recovered during data recovery, but the zigzag lines may represent routes of movement of people or goods.

It is not possible to date the execution of the various motifs at the two sites, or to determine the temporal relationship between them. The multiple motifs at AZ CC:2:348(ASM) may have resulted from an accumulation of the efforts of many individuals over a very long period of time, or a more concerted effort of a few artists over a fairly short period. The individual panels at Feature 1, for example, may be telling stories that were generated and added to over time by one or a few artists. Perhaps they were sacred localities that were returned to periodically over generations and motifs added as stories were elaborated.

## Anthropomorphs and Meandering Lines

The petroglyphs of AZ CC:2:156(ASM) are located on the south face of a small boulder and consist of an element resembling a bird track, a capital "J", and three very small linear elements. Use of curvilinear lines or forms (as represented by the "J") and connected lines to create various shapes (the "bird track") are common techniques of artistic representation.

The petroglyphs of AZ CC:2:157(ASM) are located on two boulders. Feature 1 exhibits three elements on its western face, including a geometric form resembling a Hohokam house-in-pit foundation in plan view and a possible anthropomorphic figure (Figure Thirty). Both of these designs are common to much of Greater Southwest, having been documented in a number of areas (see, for example, Bostwick 2002; Slifer 2000; Schaafsman 1980; Wallace and Holmlund 1986). Three faces of Feature 2 contain petroglyphs. Panel A (northwest face) consists of three or four very elaborate meandering lines placed closely to one another, as well as an abstract form. Panel B (southwest face) consists of an S-shaped meandering line, a solid oval, and two small pecked linear elements. These elements exhibited erosion and were difficult to discern. Panel C (northeast face) consists of a very indistinct element, possibly representing an anthropomorphic figure.

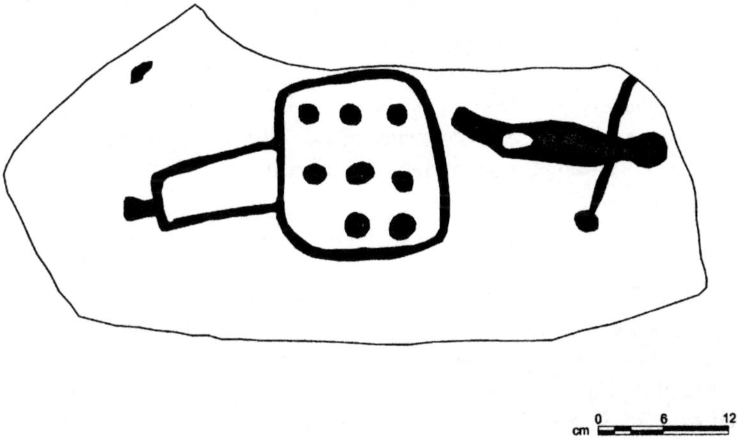


Figure 30: Feature 1, Panel A, AZ CC:2:157(ASM).

The petroglyphs of AZ CC:2:158(ASM) are located on the south face of a basalt cobble and consist of three thick, wavy or meandering lines that are set without apparent concern for their spatial associations to one another. As with the motifs described for Feature 1 at AZ CC:2:157(ASM), the specific arrangement of these forms in relation to one another is not particularly common in the Greater Southwest, yet the design itself is well-known (see, for example, Schaafsma 1980).

In general, the types of elements apparent at these three sites are different from those at AZ CC:2:231 and 348(ASM). It is interesting that the zigzag lines—the designs created by connecting small lines at their ends to create an overall linear form—so prominent at AZ CC:2:231 and 348(ASM), are wholly absent among this cluster. Overall the forms are quite distinct and might represent executions from a different time period or somewhat different cultural or ethnic affiliation. If the latter is the case, the motifs may well have had significance in terms of marking territory and/or identifying certain areas of the landscape as the possession of a particular cultural or ethnic group. One common theme at the AZ CC:2:156, 157, and 158(ASM) cluster is the elaborate meandering lines, which occur at AZ CC:2:157 and 158(ASM). These appear to be similar in technique and execution and may relate the two sites culturally, ethnically, or temporally. The "waffle iron" motif is unique, and may have served a function similar to the grid of box series present at AZ CC:2:348(ASM). Finally, the bands of wavy or meandering lines found at both AZ CC:2:157 and 158(ASM) are elements that have been defined within the Gila style of Hohokam rock art (Schaafsma 1980:Figure 60b).

The two clusters of petroglyph sites are approximately 1200 m apart over difficult terrain; whether this distance has significance in the distinctiveness of the motifs among the sites cannot be determined.

## Complex Abstract and Representational Groups

The petroglyphs of AZ CC:2:146(ASM) occur on a single boulder, with all but one element on the south face. At least six representational elements are present; three quadrupeds, two anthropomorphic figures, and one possible anthropomorphic or avian figure. A series of abstract motifs are present as well (Figure Thirty One). The quadrupeds are thick, full-bodied figures, whereas the humans and human/bird elements are stick figures. The representational motifs are very common throughout the Greater Southwest and parts of the Great Basin (see, for example, Schaafsma 1980; Slifer 2000; Wallace and Holmlund 1986). As with the spiral motif, the presentation of anthropomorphic and zoomorphic figures is consistent with the Gila style of Hohokam rock art (Schaafsma 1980:83–88).

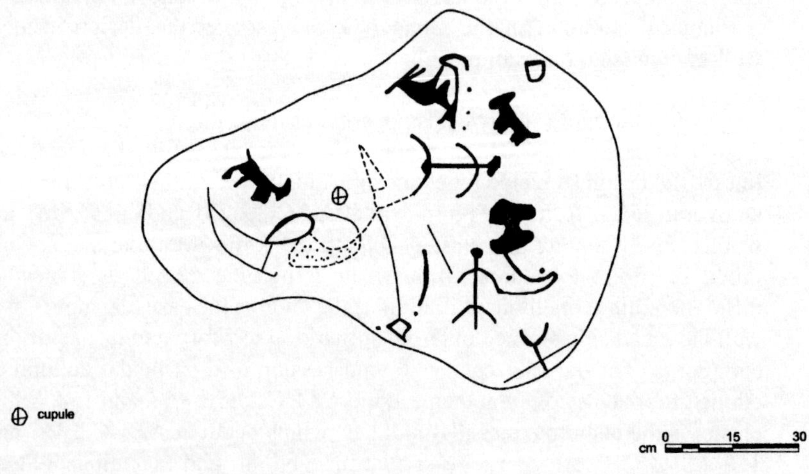


Figure 31: Feature 1, Panel A, AZ CC:2:146(ASM).

The petroglyphs of AZ CC:2:147(ASM) are located on two boulders, each with a distinct element. Feature 2 portrays a lizard-like zoomorphic figure, again consistent with the Hohokam Gila style of rock art (Figure Thirty Two), whereas Feature 3 consists of an abstract pattern created from a continuous line bent at right angles, with four small curves coming off of the exterior. There is also an area of very indistinct pecking on the face of this boulder.

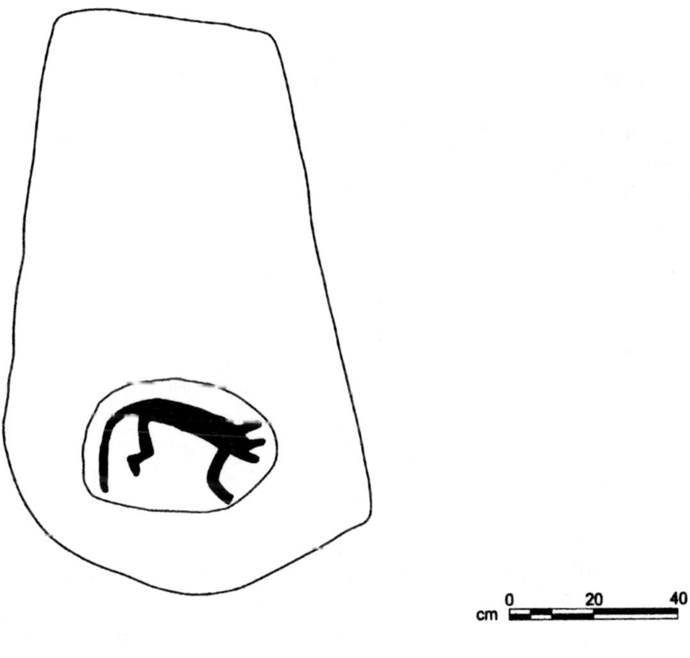


Figure 32: Feature 2, AZ CC:2:147(ASM).

The petroglyphs of AZ CC:2:149(ASM) are located on a relatively flat basalt boulder, with three petroglyph panels that are simultaneously visible when viewed from above. Elements include abstract, curvilinear, and geometric forms.

The petroglyphs of AZ CC:2:347(ASM) consists of a single motif, an elaborate spiral with a protruding segment on one side. The segment resembles a bird's head. As noted, the spiral is consistent with the Gila style of Hohokam rock art.

Among this cluster of petroglyph sites, it is possible to assign a distinct rock art style to many of the motifs, which are present at all four sites. Here is fairly clear evidence of the Gila style of Hohokam rock art, which, although not tightly dated, corresponds culturally to the later phases of the Ceramic period (between about A.D. 600 and the first European exploration of the region; see Figure Two) of southern Arizona. The range of design elements, characterized by both representational and abstract forms, is consistent more with the Tucson Basin variant of the Gila style than that of the Gila-Salt area, which tends toward a greater concentration of representational figures (Schaafsma 1980:91).

## Petroglyphs That Do Not Conform to Other Patterns

The petroglyphs of AZ CC:2:150(ASM) consists of several elements on the south face of a basalt boulder. The most obvious is an anthropomorphic figure that appears to be in motion, perhaps running. The remaining elements are faint and abstract, the boulder surface adjacent to the anthropomorphic element appears eroded. The anthropomorphic figure reflects the Gila style of Hohokam rock art.

The petroglyphs of AZ CC:2:152(ASM) consists of four basalt boulders (designated Features 1 through 4) that are located in an area approximately 200 $m^2$ on a narrow terrace above two converging drainages. Feature 1 is the largest of the boulders, and contains a complex collection of elements on both its southwest (Panel A) and northeast (Panel B) faces (Figure Thirty Three). Panel A, with at least 14 distinct elements, features one definite anthropomorphic figure, two sets of concentric circles, meandering lines, and a variety of abstract elements. Panel B contains two anthropomorphic figures, one quadruped, one possible zoomorphic figure, and at least four abstract designs. The remaining three boulders contain less distinct elements. Feature 2 exhibits a zigzag line and a variety of circles; Features 3 and 3 each contain a somewhat indistinct anthropomorphic figure. As with AZ CC:2:150(ASM), the design elements at this site reflect the Gila style of Hohokam rock art.

Figure 33: Feature 1, Panels A and B, AZ CC:2:152(ASM).

The petroglyphs of AZ CC:2:166(ASM) are contained on five small basalt boulders (designated Features 1 through 5) that overlook a drainage. A cobble feature has also been identified as part of the site, although its relationship to the petroglyphs boulders cannot be determined. Feature 1 contains panels on its southwest (Panel A) and northwest (Panel B) faces (Figure Thirty Four). Panel A contains a possible anthropomorphic figure, as well as a faint spiral and more than a dozen complex, abstract elements. Panel B consists of a line and a scroll-like motif. The remaining four boulders are pecked on one face only. The element on Feature 2 is indistinct, consisting of attached ovals; Feature 3 contains a partial spiral and a geometric element; Feature 4 has an element with a sun-like motif and an element composed of two overlapping, mirror-image designs; and Feature 5 contains a complex geometric line design. The spiral, concentric circle, and anthropomorphic figure, along with the complex juxtaposition of abstract forms at Feature 1, are all motifs common to the Gila style of Hohokam rock art.

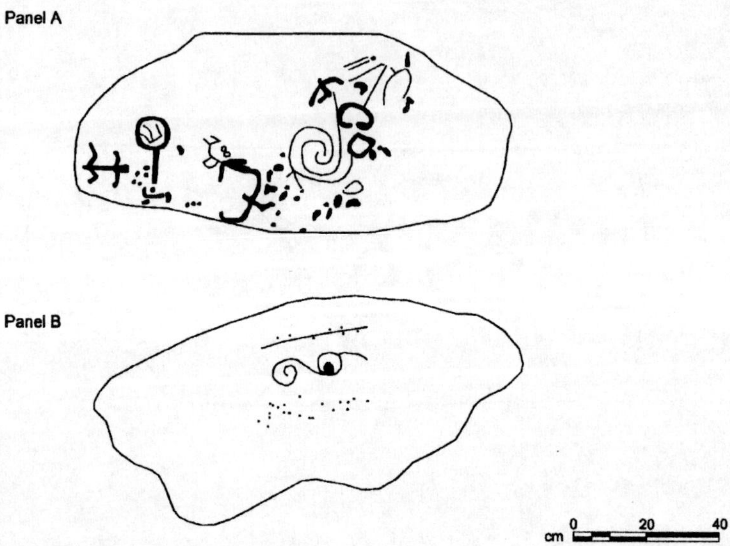


Figure 34: Feature 1, Panels A and B, AZ CC:2:166(ASM).

## Territory and Landscape as Reflected in the Dos Pobres/San Juan Project Area Petroglyphs

The only rock art style that can be confidently defined within the project area is the Hohokam Gila style. Hunter-gatherer rock art traditions, such as the Great Basin Linear style, which occurred (in regional variations) over a broad area within the Greater Southwest and Great Basin, are not in evidence (see Schaafsma 1980:33–80). Other styles, such as the art of Tom Ketchum Cave in the southeastern corner of Arizona, which fuse Hohokam and Mimbres elements, are not in evidence, either (Burton 1988:353). A few design elements, including zigzag lines and boxes with interior dots, are also part of the breadth of rock art motifs found in Chihuahua, particularly at the site of Ojos de Chuvisca (Schaafsma 1997).

There are no trails clearly associated with the rock art site, which, as noted above, occur in the western portion of the project area. Therefore, it is unlikely that they functioned as shrines, markers that identified particular routes, as Hayden (1972) has suggested for some of the petroglyphs of the Sierra Pinacate region of northern Sonora. However, the fact that they share in common elements that Hayden attributed to this function may indicate that some of the petroglyphs had meaning as "maps" or identifiers of particular features of the surrounding landscape.

The presence of spirals, concentric circles, meandering lines, and complex associations of abstract figures may point to cosmological or sacred aspects of some of the rock art in the project area. These may relate to myths dealing with origins, mythical or traditional journeys, or related phenomena. As such, the petroglyphs may have functioned as a visual cue for story-telling, for shamanic teaching, and/or for the recording of important cosmological events that created sacred space on the landscape.

The presence of representational elements—anthropomorphic and zoomorphic figures—is indicative of shamanic behavior as well. Anthropomorphic figures may represent mythological figures (once again, important in story-telling), whereas zoomorphic figures may symbolize totems or talismans, or possibly mythological figures as well. The presentation of a particular animal may have been used as a device to assist hunters in successfully finding and bringing down such creatures. Some of the human figures may represent hunters, or mythological figures who were renowned as hunters, and their presence on the landscape was a means of ensuring successful hunts. Perhaps rituals were enacted near boulders bearing such elements prior to hunting. Regardless of the specific meaning or function of anthropomorphic or zoomorphic elements, they are the visual representation of the general belief system or tradition that pertains to that landscape. Furthermore, such elements served to help define the presence of a particular cultural or ethnic group that utilized the landscape, thus providing information regarding claims of territory and rights to configure the landscape with culturally-sensitive meaning.

The prehistoric Native American cultural resources of the Dos Pobres/San Juan project area reflect infrequent and casual use of the landscape. This was a hinterland of sorts that was utilized when particular resources were desired, or as a corridor for movement between regions. Neuzil (2005b) has described the greater Safford area in the pre-contact era as a crossroads for people migrating from one region to another. Although habitation occurred and people settled into the area at certain times, the region witnessed the movement and utilization of a variety of cultures, and the scattered material culture record reflects this. Based on ceramics, Nuezil (2005:Chapter 6) has defined the presence, or at least influence, of people originating in northern Mexico (Chihuahuan), the mountain province of Arizona (Mogollon), the Mimbres and San Simon regions (Mimbres and San Simon Mogollon), the Four Corners (Zuni), and the southern desert to the east (Hohokam). Using petrographic and oxidation analyses, Neuzil (2005b:316–319) concluded that the vast majority of ceramics found in the greater Safford region after about A.D. 900 were manufactured locally, that little interregional ceramics exchange was practiced, and that this likely reflects the presence of diverse cultural groups in the region at this time. The trend for mobility and migration through the region continues into the Protohistoric period (Seymour 2006a; see Chapter Five, this volume).

Given the plethora of cultural diversity in the region during the pre-contact era, the rock art of the area served as physical markers of group identity and territory through time, a reminder of long-term ties to the land and its significance to those who used it. Perhaps lengthy periods passed when little or no use was made of these localities; regardless, human activity within the Dos Pobres/San Juan Mine project area was of short-term and limited natures. These sojourns in the foothills, carried out over several thousand years, are a reminder of the often fragile nature of the relationship between humans and the landscape in very challenging environments.

# CHAPTER FIVE

# *DESPOBLADO* OR ATHAPASKAN HEARTLAND: A METHODOLOGICAL PERSPECTIVE ON ANCESTRAL APACHE LANDSCAPE USE IN THE SAFFORD AREA

DENI J. SEYMOUR

## Introduction

Data from the Safford area bring to light questions regarding the long-held assumption that the so-called hinterlands of southern Arizona were devoid of population at first European contact, in the 1530s and 1540s (Cordell 1984; Haury 1985:396-397; Plog 1978:19; Reid and Whittlesey 1997:165). Conventional wisdom is that Athapaskan-speakers, *Dené*, or ancestral Apachean groups were not in southern Arizona until 60 to 100 years after Spanish exploration of the province of New Mexico (which included what is now Arizona). Because other non-Athapaskan mobile groups mentioned in later documents are poorly understood (Jano, Jocome, Manso, and Suma), and disappear from the historical scene in the 1700s, they are not given much consideration either during this early period.

Quite specifically, the earliest European observations recorded for this general area have suggested an absence of people in the mountains to the east and north of Safford. By implication this has also meant that the mountains and valleys south and west of there were also devoid of Athapaskan-speakers because the migration was assumed to have been from the north or perhaps the east. The Coronado Expedition (and likely Fray Marcos de Niza's) travel route has recently been identified as passing to the east of Safford (see Brasher 2007; Seymour 2007c, 2008a). Records of this expedition reference the highlands of southeastern Arizona (from the Safford area and beyond) as a *despoblado*: an uninhabited or unsettled zone (Bolton 1991:106, 108; Flint and Flint 2005:256; Hammond and Rey 1940:166; Winship 1990:16). This and other lines of reasoning have led scholars to the assumption that Athapaskans were not in central and southern Arizona until later (in the mid-1600s), when they are mentioned by name in subsequent historical documents.

The notions of an empty area and an occupational hiatus have been around for a while (Cordell 1984; Gladwin 1957:269; Haury 1985:396-397; Plog 1978:19; Reid and Whittlesey 1997:165) and embody reasoning that extends beyond the general sixteenth-century myth of an empty virgin and pristine wilderness (e.g., Bowden 1992; Nash 1967; Sauer 1971). In this traditional archaeological conceptualization of the mysterious disappearance of the ancients in the fourteenth century, depopulation by Puebloan groups, as a result of environmental factors or internecine conflict, opened a niche into which later mobile groups moved. One reason for this notion of an empty zone or an occupational hiatus is that scholars have tended to focus on Puebloan groups, accessing archaeological data that suggest geographic shifts in Puebloan occupation. Yet, new archaeological data demonstrate that mobile groups were in fact present in many areas of the mountainous Southwest before sedentary groups left, and their adaptation involved routine shifts between settlements in different locations. This might appear as a lack of occupation to observers, such as the Spaniards, many of whom practiced a settled agricultural lifeway, when in fact, people in the mountainous portions of the Southwest were just at a different settlement, using an alternate resource zone, or in some other topographic setting. The *despoblado* of the early European chronicles is therefore suggested to be more apparent than real. This early presence leaves open the possibility that mobile groups were responsible for clearing out the resident sedentary populations or that they were at least present for the reorganization that resulted in aggregation of Puebloan groups into fewer locations. This concept of the Athapaskans "illegitimately" interfering with the Puebloan presence is a notion that has not been popular for some time (see Gladwin 1957:267; Kidder 1975:341).

This paper examines the notion of an empty area in the mountainous heart of the American Southwest. This discussion questions the premise that

Athapaskans migrated into an empty area long after it was abandoned by prehistoric populations and that the mountainous area in the vicinity of Safford was devoid of population even as late as the Coronado Expedition in 1540. What follows is not a "could have, so must have" type of argument, but rather, archaeological data contradict the documentary record and therefore it becomes necessary to reinterpret the later. Ample and diverse archaeological data including early dates directly associated with Athapaskan material culture point to a widespread early Athapaskan presence, but these data are often precluded from consideration, dismissed out of hand, because of the undue precedence given documentary sources. Because the documentary record has played such an important role in shaping perceptions it is important to consider the nature of that record and the methodologies employed in its interpretation. It is also necessary to explore the cultural and historical context for the notion of a *despoblado* before proceeding on to a summary of the more substantive data of the archaeological record that places mobile groups in the area in the pre-Coronado era.

## Some Methodology-based Implications

An interesting methodological issue is raised by the differences in conclusions reached on the basis of whether emphasis is placed on documentary versus archaeological data sources. Even when appropriate methodologies have been employed in the use of documentary evidence an inaccurate conclusion has arisen. The inappropriateness of inferences drawn from the texts is now evident only because an alternate, independent, and objective data source has been used and because the textual content has been independently assessed and found to be misleading.

## Seriating Sources and Implications of Routes Taken

Ethnohistorians and archaeologists who use historical data understand the importance of temporally seriating documentary sources so that inferences regarding past lifeways are not collapsed in a way that normalizes the record and prohibits recognition of change. When sixteenth and early seventeenth century documentary sources are seriated a seemingly straightforward impression is gained that begins with a lack of notice of mobile groups in the area known later as their heartland. Shortly thereafter mobile groups are consistently seen by visitors at the fringe of this upland heartland. In this conventional scenario, it stands to reason that (1) such groups should have focused on this heartland area from their earliest presence and so, (2) since mobile groups were not reported during this leg of Coronado's journey, no such groups were present at that time. The logical inference is that mobile groups were not present until noted in the later set of documents. The Coronado Expedition then becomes a time marker: no Athapaskans are expected to be present before this expedition. As Reid and Whittlesey (1997:165) note:

> The mountains of central Arizona were completely abandoned when Francisco Vásquez de Coronado marched his vanguard through them in the early summer of 1540, forming a great *despoblado* that was perilous for man and beast.

Yet, these documentary-derived inferences about an empty wilderness are incorrect. This only becomes apparent when sources external to the documents are examined that prompt one to consider the circumstances of discovery.

One of the key variables affecting the observance of mobile groups by early explorers was the routes taken and instances of variance from those valley bottom routes. Even when mobile groups are observed 40 years later they are seen only when Spaniards ventured off the river valley pathways into the foothills bordering the rugged mountains. These earliest mobile groups were noted in the mountainous region of eastern Arizona and western New Mexico at the edge of Mimbres territory or in the Socorro region. In 1581 the Chamuscado-Rodríguez Expedition, named for Captain Francisco Sánchez Chamuscado, the military leader of the group, and the religious head of the group, Fray Agustín Rodríguez, traveled through an uninhabited stretch for 15 days and 70 leagues north from El Paso along the Rio Grande. It was not until leaving the valley to explore a foothills trail, after going more than two leagues, that:

> On entering the mountainous territory, we saw an Indian brave and two inhabited huts. Taking our horses and arms we went in that direction and discovered many people, who fled toward the mountains when they saw us approaching (Hammond and Rey 1966:80-81).

Hammond and Rey (1966:80, Footnote 1) suggest that these 70 leagues brought them to just south of San Marcial.

The Espejo Expedition experienced this Rio Grande *despoblado* in 1582;

they did not encounter inhabitants until 15 days north of the El Paso area. After traversing this uninhabited corridor along the Rio Grande, Antonio de Espejo, a wealthy citizen and expedition leader, noted:

> At the end of this march, we came upon a thinly populated ranchería containing a number of straw huts. Here we found many deerskins as well dressed as those brought from Flanders, quantities of excellent white salt, some jerked venison, and other provisions. The Indians of the ranchería welcomed us . . . (Hammond and Rey 1966:1966:219).

This again was probably just south of San Marcial, which is the first location north of the Robledo and Uves mountains of the El Paso vicinity where a series of hills extend east toward the river providing suitable terrain characteristics for mobile group occupation. The lesson to be taken from this is that a differential distribution in settlements does not mean the area was generally uninhabited, only that certain segments of the Spaniard's trail were devoid of occupation. It is only when the visitors left the river corridor or valley bottom that they encountered mobile groups, and when they came across mobile groups they were at the fringe of the area in which Coronado found his apparent *despoblado*.

It is also relevant that the extended uninhabited stretch along the Rio Grande is situated where canyon walls are steep and high, restricting access to the river bottom, therefore making the area less attractive for riverside occupation by agriculturalists. Agricultural settlements began just to the north of the "straw huts" where topographic features were more conducive to river access and farming. Again, the absence of people in this narrow corridor to the south does not mean the entire region was devoid of population, just as the absence of people along Coronado's route does not imply this. Though researchers have extended the lack of native use of Coronado's restricted travel corridor to mean no ancestral Apache were present at all in the sense that they had not yet migrated into the area, this is simply not the case as external sources indicate and as contextual analysis explains. Known ancestral Apache sites are perched in the mountainous terrain along, bordering, and overlooking the Coronado route.

Diego Pérez de Luxan, a member of the Espejo Expedition, writes of this same expedition, but farther south, that columns of smoke were seen high in the sierra on the opposite side of the river (Hammond and Rey 1929:71). In Bolton's (1916:176, Footnote 4) interpretation the expedition had crossed the river, which would mean these smoke columns were rising from the Mimbres Mountains to the west of the river, but as Hammond and Rey (1929:71, Footnote 49) point out there is no basis for suggesting the travelers crossed the river and so these columns may have been east of the river. In any event, this provides evidence of human occupation, in all likelihood by mobile groups who tended to utilize such zones, in mountains that are even farther to the south as early as 1582.

By the early seventeenth century a conclusion to be inferred from the historic record is that the mountainous areas adjacent to the Gila River in New Mexico and Arizona formed the homeland or the core of an important early Athapaskan (ancestral Apachean) occupation. Thus, within 80 years of the Coronado Expedition the uplands above the Mogollon Rim that constitute at least a portion of the referenced *despoblado* were recognized as the very heartland of occupation by certain ancestral Apachean groups. Its occupants were referred to as the Gileño or Gila Apache. In the 1620s, Benavides mentioned the Province of the Apaches of Xila (Gila), situated in Socorro County, New Mexico about 40 miles west of the Rio Grande. He further mentioned that the principal Apachean chief spread the gospel to his rancherías—indicating that other settlements were present (Ayer 1965:42). He specifically noted that going northward from there the area was "full of rancherías of the jurisdiction of Xila" (Ayer 1965:44).

Other documentary evidence supports this notion of Athapaskan groups being present along the Gila and the bordering mountains, though a bit later in time. Specifically, in 1667 the Mansos (a group of unknown affiliation, but see Seymour 2002, 2004a, 2008e) of the El Paso area were in collusion with the Apaches under the headship of the Gileño leader El Chilmo. This information was recorded because some Manso rejected missionization and chose the mobile rebellious lifeway (Forbes 1960:162).

Through time these areas along the Gila and the bordering mountains that lie within Coronado's *despoblado* became known even more widely as the "center of the Apache nation" (Naylor and Polzer 1986:652, also see 651). Spanish documents from the 1690s translated by Naylor and Polzer (1986) express the frustration and generally held knowledge by the presidial soldiers that the Apache and their allies retreated to and found unmolested refuge in the Gila River area (Naylor and Polzer 1986:652, also see 651). During a summer campaign of 1695 Captain Juan Fernández de la Fuente reported that the "…the Janos, Jocomes, [..ovias?], Mansos, Chinarras, and many Apaches were assembled in the woods along the Gila River" (Naylor and Polzer 1986:645; also see Naylor and Polzer 1986:486, 584, 641-645, 648). Indian captives reported "The Janos Sumas, Mansos, Chinarras, and Apaches …were to be found in the woods on the Gila River and in the marsh and arroyos that rise on the north side of the Santa Rosa Mountains [Pinaleño Mountains] in the home of the Apache" (Naylor and Polzer 1986:648). More specifically, "The Janos were located at the spring on the northern slope of the Santa Rosa Mountains, whereas the Jocomes and Apaches were located on the west side of the Santa Rosa Mountains" (Naylor and Polzer 1986:644).

My work in this area and evaluation of the work of other scholars suggests that the documentary-record-derived assumption of an early empty wilderness has been too broadly extended, and the cultural and historical basis for this misconception can be traced. Early excursions crossed by and near the mountain ranges that surround Safford and the Gila Valley, and surround the apparent *despoblado* area where mobile groups were observed 40 years after Coronado. These sightings do not embody an accurate representation of the distributions of these mobile groups at the time, but, rather, these later viewings attest to the widespread presence of mobile groups in areas surrounding the *despoblado*. The historic sightings are from all around the periphery of the mountains that define the Coronado *despoblado,* limited in extent only by the routes Spaniards followed (Figure Thirty Five). This presents the possibility, confirmed by other data sources, that the mountains were occupied by, if not teaming with, mobile populations, some of whom may have been Athapaskan-speakers.

The route chosen by Coronado was along a well-established trade route and trail to Cibola (see Brasher 2007; Seymour 2007c, 2008a). It is unlikely that mobile groups would position their sites on a trail that was frequented by foreigners that came from as far away as West Mexico. Rather, they would situate their settlements so as to have a view of the route, while remaining safely at a distance. Many proto-Apache sites in southeastern Arizona and western New Mexico occupy such settings, overlooking but at a distance from trails, prompting suspicion that they knew in advance when foreigners were on their way and could easily avoid notice.

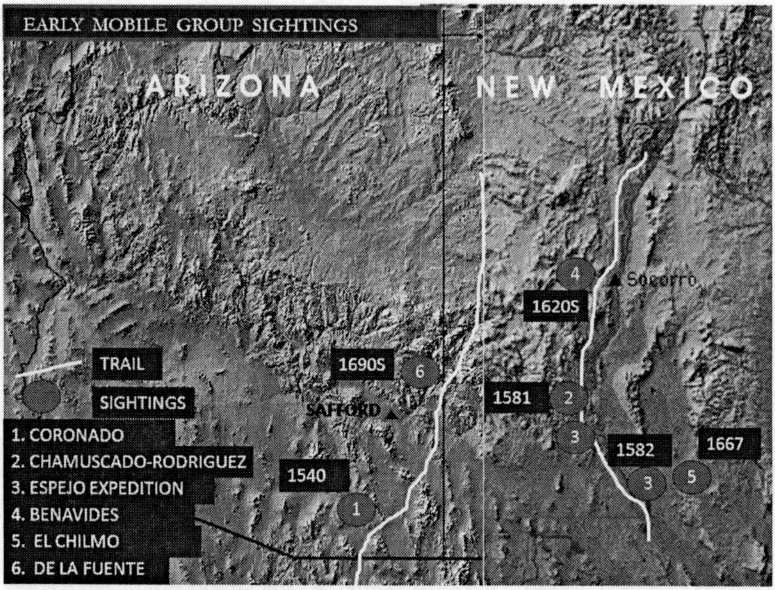


Figure 35: Historical sightings of mobile people near the Mogollon Highlands in the Spanish Period. Dots follow frequently traveled trails.

A portion of the Coronado route through the *despoblado* involved traversing mountains and dangerous passes and moving through a piñon and oak forest (Flint and Flint 2005:256; 417, 473; Hammond and Rey 1940:251). Much of their travels through the *despoblado*, however, likely included the grasslands in the valley bottoms, for Castañeda noted (Flint and Flint 2005:417) that after Chichilticali "one finally crosses the mountainous land, [which] is broken to permit passage to the land's region of plains." The terrain throughout this area is varied and includes expansive grasslands bordered by rugged tree-strewn mountains. These valley-bottom grasslands or plains were not the favored terrain for Athapaskan habitation (though they did use low-lying areas) and thus were not likely the focus of Athapaskan resource use or settlement. Consequently, this area would have been unsettled. To the Spaniards, who did likely traverse these grasslands owing to the ease of travel, this absence here would likely suggest a more widespread absence of people, especially since it was in lowland settings that the Spaniards had encountered most of the other groups along the way. Yet, suitable portions of the terrain between the thorny-treed desert described in the Coronado accounts and Cibola contain archaeological evidence of an early Athapaskan presence in the mountains that border the river valleys.

It is suggested that mobile people were later viewed in these limited areas because these were the few circumstances where the Europeans left the river corridors, where the natives left the mountains, or where their presence could be discerned in the mountains from the distant valleys. The early Athapaskans in the Southwest who arrived via a western series of corridors, like their later Apachean descendants, were a mountain-adapted people and though they used the valleys they retreated to the mountains for safety. This not only makes it likely that they inhabited the mountainous terrain surrounding the so-called *despoblado* at the time of Coronado, but it also probably accounts for why the Spaniards did not see them (Danson 1957:116; Harrington 1940:510; Seymour 2002, 2004a:186, 2004b). Such groups were most easily viewed at the periphery of these steep and rugged ranges, not when in the deep woods or rugged rocky mountainous slopes. This likely accounts for their apparent absence from the region, whereas in fact, the ancestral Apache were probably present and inhabiting the nearby rocky terrain and mountains.

It is also likely that some of the native groups hid, as they did from a 1695 Spanish military campaign: "When the two in question observed so many soldiers, they became frightened, took flight, and hid" (Naylor and Polzer 1986:645). Moreover, later presidial soldiers did not always have luck finding sign of these groups even when in areas where they were known to be present (see for example, Naylor and Polzer 1986:651). The extended nature and many failures of the nineteenth century Apache Wars also attest to this invisibility of Apachean groups, which was part of their defensive strategy. Even Chiricahua legend notes that their ancestors successfully hid from the Spaniards for the lifetime of one man (Cole 1988) suggesting that they were aware of the presence of Spaniards before the Spaniards were aware of them and that hiding was a sound strategy from ancient times.

Thus while seriating documentary sources is a time honored technique, it is imperative to also consider operative factors that lie outside the documents. Contextualization of the texts in a matrix of archaeological data, situated in a wider geographic framework, and grounded in an understanding of mobile group landscape use presents the basis for a different perspective. A deeper time perspective—over the *longue durée*—that incorporates archaeological evidence, reveals that some of our basic assumptions regarding the dawn of history in this area are in need of revision.

## Crosschecking Sources

Another premise of documentary analysis is that sources are incomplete, potentially biased, and representative of the European perspective and that they cannot be taken at face value. In fact archaeologists have been criticized for assuming that documents can be used in a straightforward way and that we do not understand the difference between what is actually being said and how it is articulated (Galloway 1991:455). Ethnohistorians attempt to surmount this problem by crosschecking sources, which Dobyns (1966:403) referred to as the "simplest empirical method of source evaluation." This generally accepted method of assessing reliability and even the validity of documentary data involves using a number of parallel sources so that texts of a variety of types can be compared and contrasted to provide perspective and to ensure consistency.

Yet, when crosschecking if all versions of a subject derive from the same observations made along the same path that bypassed mobile group settlements and resource use areas (that were distributed on the basis of certain spatially restricted topographic characteristics), simply increasing the number of texts examined may not provide an accurate representation. The insidiousness is that there may be no serious contradictions to expose internal flaws. As should be apparent by now, what is needed is an independent form of verification that is external to the expedition.

## Contextualizing Sources: The *Despoblado* in Cultural and Historical Context

Ethnohistorians also attack this problem of understanding the role of a text by contextualizing the document. This source criticism can expose potential objectives and other forces behind text content. Archaeologists trained in history or ethnohistory often expect contextualization to proceed within a traditional historical format, which requires one to recount the basic descriptive and analytical conclusions of others that surround the topic, whether relevant to the discussion or not. This is now widely considered an invalid expectation. It is a holdover to what Harlan (2007:507) has discussed as the cult of immersion and is attributable to "the apparent belief of some historians that immersion means first reading all prior work in the topical area and then examining each textual source with equal intensity." As Harlan (2007:508) continues

> Archaeologists who utilize documents have a tendency to violate most, if not all, of the historian's culturally based precepts. At the very beginning of an inquiry, they are quite willing to use translations when the research in question requires no internal source criticism.

Rather than redundantly recounting an obligatory set of expected, though marginally relevant factors, the archaeology itself provides a reliable form of contextualization that is valid on its own, often making consideration of other contexts irrelevant and unnecessary (especially since they take up valuable space in an article). After all, a reciprocal requirement is not standard procedure; few would accept the argument that historians and ethnohistorians should not proceed with contextual analysis in the absence of objective and externally confirming archaeological data.

By placing the Coronado Expedition within its cultural and historical contexts it is possible to infer that the *despoblado* may not be as it appears from a solely textual point of view. In order to understand the basis for this notion of a *despoblado* several cultural and historical factors must be considered that relate both to the Spaniards and the native occupants. These include two types of issues: (1) perceptions and motivations of the Europeans and (2) native landscape use and terrain selection that affected the ability of the Spaniards to see the mobile groups or their ability to remain hidden from the Spaniards.

*Perceptions and Motivations of the Europeans*. To reiterate, this perception of an empty wilderness in the American Southwest largely derives from the Coronado Expedition (1540-1542), which passed through the general Safford area (see Seymour 2008a). As noted, its chroniclers claimed that an extensive part of the terrain there and to the northeast was a *despoblado* or uninhabited zone, a perception that stemmed from the lack of encounter between expedition members and natives during a 12- to 15-day period over 80-leagues (200-240 miles; Hammond and Rey 1940:198). Local natives also expressed this view; they either shared the perception that the area was unsettled or conveyed a kind of information to the Spaniards that was used to infer this. This latter point is important because "unsettled" versus "uninhabited" convey very different connotations when one considers the "unsettled" adaptations of the groups present. The imprecise nature of word use in Spanish means that *despoblado* can be translated in any number of ways, as it has been, for example: unsettled (Flint and Flint 2005:256), uninhabited (Bolton 1991:106, 108; Hammond and Rey 1940:166), and wilderness (Winship 1990:16). While "uninhabited" is fairly straightforward, "unsettled" may mean that mobile or wandering people were present, but settled or sedentary people were not. Wilderness may be interpreted a variety of ways, including that "wild" or "untamed" people, as these populations were often viewed, were present who did not modify their environment through city building and agriculture, but rather were seen as part of the wildness.

As I have pointed out elsewhere, these terms and perceptions seem to be burdened with numerous culturally, economically, and politically charged connotations (Seymour 2008a). Specifically, when at Chichilticali at the edge of the *despoblado* mention is made of encountering people that were referred to as barbarous [*mas barbara*] or uncivilized. The political and economic implications of Castañeda's reference become apparent when considering sixteenth-century conceptions of property rights (Pagden 1990) and the goals of expedition members to lay claim to and ultimately settle their discoveries (Flint and Flint 2005). Legal discussions in Spanish society in the early 1500s began to focus on the issue of the property rights of conquered peoples (Pagden 1990). These issues included discussion of the right of conquerors to claim land and minerals, as well as to harness native labor. Questions of the disposition of discovered property gained relevance in the 1530s because of lack of clarity along these lines and because of the important economic implications of the answer. Roman law was invoked to address these issues, using the logic that barbarous or primitive inhabitants lived without civil society that was based upon real property ownership. The absence of property relationships among mobile groups, meant that they and their social structure was not civil, and therefore their members could not make claim to the lands conceived as open spaces or wilderness that they happened to inhabit (Pagden 1990:15). Thus the barbarous or savage natives might wanderer throughout open lands, but because they did not settle, exploit, or control the land, the natives had neglected the land, and therefore forfeited rights to it, making the land available to the conqueror and Crown (Pagden 1990:15-17).

Though some Spaniards may have been aware of nomadic populations and their wandering adaptation as such groups were encountered during campaigns in northern Africa of the late fifteenth and early sixteenth centuries, this does not seem to have affected their understanding of the lands through which they passed. One reason for this likely relates to the fact that conquest of North Africa at this time focused on defeat and possession of coastal Muslim-populated towns in the war for the faith against the Moors. The hinterlands remained unoccupied and under local control and therefore outside the view of the invaders (Elliott 2002:53-54; Kamen 2003:30-33). Thus, it is easy to see how perceptions regarding mobility may very well have contributed to the wording of these historical passages.

*The Nature of Athapaskan Sites, Landscape Use, and Terrain Selection.* To understand issues surrounding the *despoblado* it is useful to consider certain key concepts relevant to mobile group landscape use and terrain choices. Mobile groups tend to select residential locations on the basis of the distinctive characteristics or the performance characteristics intrinsic to a place. Rather than making, modifying, or structuring a place upon arrival, locations are sought that naturally possess relevant terrain and topographic features (Seymour 2002, 2005, 2007d, 2009). This proclivity to seek out locations that are already suitable for predicted and priority needs relates to the short duration of expected stay. When highly mobile, a group will look at the appropriateness of a location rather than expending great amounts of effort in modifying a place to suit their needs. Thus, mobile groups tend to seek out rocky locations where they can easily gather rocks together to create the foundation for their structures, or better yet, they will move a couple of rocks into place against an already suitable outcrop. They tend to locate in places with appropriate materials for making the superstructure and they camp in proximity to water, fuel wood, and other basics that can be easily accessed. Mobility allows greater freedom to choose from a variety of places that possess various combinations of necessary space, natural site accoutrements, and resources.

When mobile groups are at the fringe of the territories of more sedentary groups or of nations that are hostile to them, they also seek out locations that are defensible, that have expansive views while not exposing themselves to discovery, that provide materials and camouflage for shelters, that allow frolicking without concern of detection, and that permit direct line-of-sight to lookout stations and to places of danger. Many times these security and safety measures are given greater weight in the selection process at the expense of convenience. For example, usually such sites will be found at some distance from water (Seymour 2002). This is because water is transportable to a safe location, whereas "others" and enemies know of the limited water sources and so sources become magnets for unexpected visitation and are predictable way stations for most travelers. By situating away from accessible water sources a group is less likely to be surprised. The same logic was applied to proximity of encampments to widely used trails. In remote areas that are deep within a buffered territory or far from concern of intruders, encampments are found nearer water sources. The cloak of safety provided by remoteness is likely the reason for these distributions of sites nearer water, explaining why some Apache groups were occasionally surprised by the Spanish military, who apparently got lucky in finding them in these lower elevation settings by foraying into remote zones were they were not expected.

Athapaskan site distributions change with respect to natural water sources (springs, washes, seeps, etc.) through time because of their relationship with other groups (Seymour 2002). Early on and later in remote areas, they tended to camp in small groups near these sources, including along the major rivers (Rio Grande, San Pedro). Later with European pressures, and also in accessible areas, they camped at least one-quarter to one-half mile from springs so as not to be surprised (Seymour 1995, 2002). What is most interesting, however, is that some major habitation sites are not anywhere near springs or streams. For example, a Peloncillo Mountain site (AZ CC:12:58 [ASM]) east of Safford is 12 miles from the nearest known spring. The logical question relates to where they got their water. Their solutions differed from those devised by more settled populations because they had a different mindset about water, how to extract it, and what was sufficient. Many locations where they settled presented just a trickle or they used what collected in natural basins. It is common to find natural huecos near most habitation sites in the Hueco Bolson and Tularosa Basin, suggesting that this was a selection criterion (Seymour 2002). Natural crevices and mortar cups served as collection basins at sites in the Peloncillos, along with specialized water collection features that were constructed of curvilinear arrangement of rocks on flat shelves in crevices and overhangs below natural seeps.

A practical consideration for widely dispersed mobile groups is that they must be able to find one another. This becomes a concern either because associates disperse without notice when under attack or because being dispersed much of the year they gather seasonally or for particular purposes in large groups from many far-reaching areas. Thus, notable features of the landscape allow participants to distinguish with "rich descriptive imagery" between a diverse range of possible residential meeting areas, that are given "handsomely crafted names—bold, visual, evocative" (see Basso 1996:23). Historical references note that decisions were always made in advance as to where people would regroup if they became dispersed after an attack (Ball 1970:77, 79, 173, 178; Betzinez and Nye 1959:131; Quaife 1935:132; Seymour 1995, 2002). These notable landscape features may include a prominent peak, a rocky projection, a colored (white or red) inclusion or dike or stratum in an otherwise brown or blue mountain range. Even as late as the 1860s rancherías were situated on "a bold bluff standing prominently out into the valley, and commanding a view of every possible approach from the valley" (Sweeney 1992:25). This pre-planning tactic that was focused on prominent features of the landscape allowed people to find one another while remaining completely outside the awareness of the enemy. Use of a relatively consistent set of "trails" between serially used encampments facilitated encounter with friendlies and avoidance of hostiles.

Understanding these relevant landscape-oriented factors furnishes the necessary background for understanding the apparent *despoblado*. Residential sites will not be evenly distributed across the landscape because of (1) the tendency to use mountains rather than valleys, (2) operation within a distinctive and restricted landscape sphere, (3) attraction to certain topographic settings, and (4) selection of specific terrain features. One valley may contain no residential sites, whereas groups may be present in abundance in the adjacent mountains (Figure Thirty Six). Even in that mountain range, most of the canyons may be devoid of occupation whereas those that possess relevant distinctive characteristics will be occupied. A canyon with a backdoor escape and boulder-studded sides will be the focus of residential use, whereas a shallow canyon with barren slopes might not (Seymour 1995, 2002).

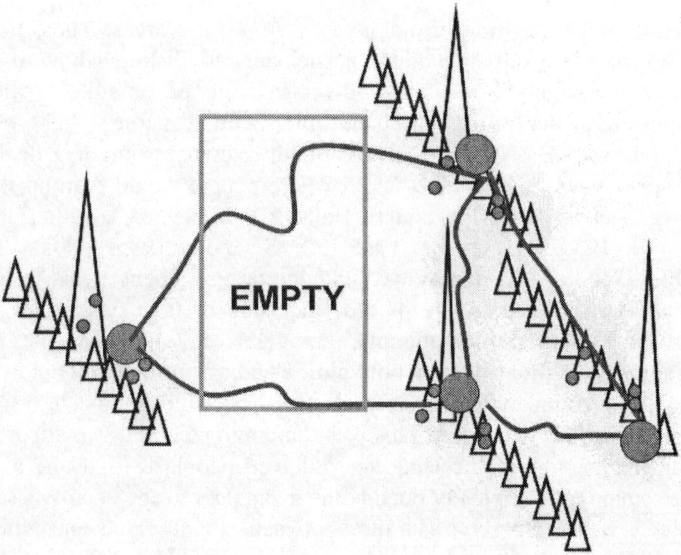


Figure 36: Simplified schematic model of Apache site distributions and landscape use. Circles represent ancestral Apachean sites in mountainous settings linked by trails across the unoccupied basin in between.

Encampments will also be clustered in these favored areas. The clustering sometimes results from visitation to the same general location by the same group many years running, so that each time they visit they establish an encampment in a new location. Equally often, however, different bands converged in these locations at the same or different times, each inhabiting their own sector of a mountain, ridge, or canyon, but all being attracted by the performance characteristics of that particular place. This is consistent with the ranchería-style of landscape use that is so typical of these mobile groups (Seymour 2002).

Because locations of large sites are keyed to prominent landforms their placement is relatively predictable, for the participants and for the archaeologist (Seymour 2002). Because high-elevation settings were selected, views from these settlements are expansive. Consequently, these large sites are often inter-visible and occupants could easily signal one another or would be immediately aware of one another's peril if in fact settlements in different ranges were occupied at the same time (Figure

Thirty Seven). Usually though, a group established settlements in several different mountain ranges that could be visited at different times by the same group. When looking for one another without the benefit of a pre-planned rendezvous such inter-visibility would allow individuals to reunite with the group. This use of bases in several different ranges meant sensible conservation of resources but it also had the added benefit of confusing pursuers as to which mountain range a group tended to occupy. These high-elevation settlements also permitted clandestine observation of oblivious basin-bottom travelers (such as the Coronado Expedition) for days in advance.

Mobile group site distributions are all about terrain selection and landscape use, which in turn are related to relationships with surrounding groups who have laid claim to particular tracts of land. The template for Athapaskan settlement selection is diametrically opposed to our own perceptions. In our view the mountains are wilderness and the valleys are habitation-friendly locations, whereas Apachean groups at conquest and after viewed it in exactly the opposite way. The mountains were their sanctuary while the valleys brought contact with other already established groups who were sometimes hostile to mobile interluders. Thus, important settlements are expected most often in and near the mountains rather than in the open valleys. These terrain preference and landscape selection criteria do not mean a region was not occupied, it simply means that only certain niches were occupied. In the case of the Mogollon uplands the route through which Coronado traveled apparently did not intersect with the Athapaskan niche and so the area at large seemed unoccupied.

Another factor affecting site distributions, landscape use, and visibility to historic trekkers is that many mobile groups, including the early Athapaskans seem to have practiced a form of seasonal transhumance. In the winter the ancestral southern Apache seem to have occupied small campsites at low elevations. Relatively small families or local groups spent the cooler season on south-facing or east-facing slopes, hidden in canyon mouths, in round valleys, and at the confluence of washes. These small sites were sometimes occupied repeatedly on an annual basis, building a relatively rich record, with tens of artifacts and small roasting pits that show evidence of reuse. For Apache sites, tens of artifacts are considered high density because on small sites and especially on sites occupied in the nineteenth century artifacts are scarce. Frequencies of artifacts of this "magnitude" suggest long-term or repeated used of a site, particularly if diversity is high. For example, an early site in the Dragoon Mountains of southern Arizona that dates sometime between the A.D. 1300s and 1500s consists of two structures, two roasting pits, two caches, a gaming circle feature, but only 18 artifacts in the core of the site.

Figure 37: Inter-visibility of sites and prominence of landforms.

Sometimes larger groups occupied the lower-elevation areas. At the early end of the Athapaskan occupational sequence in the southern Southwest these lower elevation settings were selected near springs and in canyon mouths. These foothill and canyon areas were relatively safe before the arrival of Europeans and would have been favorable for settlement before the migrants became familiar with the intricacies of their landscape (Seymour 2004b).

In other circumstances, the process of small independent groups camping in the same general location, but moving their camp across the slope or around the ridge finger, led to a build-up of numerous structures in low-elevation settings. The result of this accumulation of episodic settlements is the formation of a large archaeological site with numerous structural features, but low densities and perhaps low diversities of debris. A large group on the move, where they built structures for overnight use, but then moved on to another location, might create the same settlement pattern, site layout, and material culture densities as these smaller accumulations. In later historic times, this type of en masse movement tended to occur when moving to a new encampment in front of an enemy force that could be seen advancing along the valley bottoms from the high peaks for days in advance (Betzinez and Nye 1959:65), though closer encounters would result in dispersion.

Ancestral Mescalero and Chiricahua Apache groups also occupied larger camps (Seymour 2002, 2004a). Until recently nothing was known of these larger settlements but now I have documented four of these in the Whitlock, Peloncillo, Hueco, and Alamo mountains. The early historic record indicates that these were sometimes occupied by multiple bands, consisting of thousands of people. When occupied by a single band in the nineteenth century these habitation sites sometimes contained 100 wickiups, sheltering 500 people, 75 of whom might be warriors (Ball 1970; Betzinez and Nye 1959). Similarly, historic records indicate that eighteenth-century rancherías varied in size, sometimes becoming quite large: "There are some rancherías of eighty and one hundred families or of forty, twenty, or less" (John 1989:62). Nineteenth-century reports are known of camps with two and three hundred people (Sweeney 1992:27). Those I have documented have hundreds of structures, as well as other feature types indicating that these safe and favored locations were the focus of repeated use for short periods by several distinct bands over many centuries (Seymour 2002, 2004b).

These large sites tended to be situated deep in the heart of the mountains, where occupants were safe from surprise intrusions, or far beyond the normal routes of military campaigns. As of 1799, José Cortés, Lieutenant in the Royal Corps of Engineers (John 1989:65), reported several reasons why independent groups from several rancherías might temporarily coalesce into large rancherías:

> By chance it happens that sometimes many *rancherías* come together at one place in search of certain fruits that abound in one given location or another. Others are gathered together by prior plan or design with the notion of forming a body for their defense or with the idea of holding one of their ceremonies or where they gather for the hunt. At these ceremonies they often decide upon some plan to attack their enemies. In such instances not only do the *rancherías* of one group combine, but often two

> or more whole tribes will congregate.

These large residential sites were often at high elevations (Betzinez and Nye 1959). Sources refer to the hilltop or high-elevation villages occupied by numerous Apache bands (Ball 1970:22; Betzinez and Nye 1959:85; Cole 1988; Sweeney 1991, 1992). Again Cortés (John 1989:65) noted:

> Thus gathered, the rancherías always occupy the steepest canyons in the mountains, surrounded by the most difficult passes for approaching the site where they are located. The site is chosen, as a general rule, adjacent to the greatest heights in order to command the surrounding valleys and plains.

These landscape-use and terrain-selection criteria are crucial for understanding Coronado's perceived *despoblado*. Ancestral Apachean focus on these upland zones also explains why archaeologists who focus on Puebloan sites near low-lying arable land have generally been unsuccessful in finding Athapaskan sites.

## Subordination of Archaeological Sources

Because of the long-held practice of treating non-textual sources as secondary and supplemental, the documentary record has been used as the basis for inferences about the timing of arrival of Athapaskans into the southern Southwest. This arises because archaeology is seen as a deficient source whose content is substantially limited, and as Harlan (2007:508) notes, some historians believe "that written documents are the only true source of information about the past." Because of this, archaeology is called upon to support the documentary record or, alternatively, documentary and ethnographic sources are used to help interpret ambiguous archaeological results (Seymour 2007a).

This tendency to subordinate the archaeological record to the documentary has meant that archaeology is not viewed as an independent source but as a supportive one. Thus archaeology is seen as less complete than the documentary record and dependent upon it for interpretation, whereas in fact, archaeology is a distinct and independent line of evidence. Because of this independence, use of both sources together does not lead to circular arguments, as is sometimes reasoned. While both of these source materials may refer to the same temporal and spatial subject they are independently constituted and in no way does one source depend upon or arise from the other (Lightfoot 1995). Nor is either an exact replication of the subject, but instead both are distorted representations of past behavioral systems. This is what makes their combined use so productive and attractive and what makes this approach so different from the direct historical approach of the early twentieth century (see Lyman and O'Brian 2001).

To counter this inaccurate documentary-based perception, archaeological data are enlisted to document that early Athapaskan sites are present in this mountainous region and these predate the earliest European expeditions. They are also present further south before European presence, suggesting that the migration through and settlement of the general region of the *despoblado* had already transpired. Information regarding these sites and chronometric dates from them are available to substantiate these assertions (Seymour 2002, 2003a, 2003b, 2005).

By adopting a wider perspective that incorporates objective data from the archaeological record it is possible to see that the Coronado documents are misleading regarding this issue of the *despoblado* and the timing of proto-Apachean occupation in the Mogollon uplands. These supplemental data derive not from one site but from dozens, though the earliest dates that fall solely in the pre-1540 period are so far limited to nine sites (Table Six). These conclusions are based on not one date but from date suites from single sites and from across a number of sites that, as we shall see, also contain distinctively Athapaskan material culture (Seymour 2005).

Moreover, until recently the best-dated Athapaskan sites in this mountainous area (for example, the Grasshopper Spring Site, Reid 1997b) have all dated to the 1600s, providing archaeological data that would seem to support this tidy documentary-derived scenario. But the assumption of an empty zone has meant that until recently scholars have not earnestly looked for evidence of early Athapaskans, because they have not expected them to be there. The unobtrusive archaeological remains relating to Athapaskans and other mobile groups will not be seen unless specifically and resolutely searched for, in part because the early pre-differentiation record is far fainter than later Apachean evidence. Recently obtained archaeological data indicate that there is no empty niche, that Athapaskans were present, and that the negative evidence regarding this historically referenced group has been endowed with more meaning than is warranted.

**Table 6: Athapaskan Chronometric Dates from the Southern Southwest that Predate Coronado**

| DATE | TECHNIQUE | SITE |
|---|---|---|
| A.D. 1270 to 1320 (680 to 630 Cal B.P.) and Cal A.D. 1350 to 1390 (600 to 560 Cal B.P.) (Beta-203376) | C14 (Calibrated, 2-sigma) | AZ CC:7:11[BLM], Whitlock Mountains, AZ, grass from storage platform |
| A.D. 1415 to 1490 (535 to 460 Cal B.P.) (Beta-188760) | C14 (Calibrated, 2-sigma) | LA 117112/FB 16265, Rough Canyon, NM, grass from storage platform |
| *A.D. 1450 to 1670 and A.D. 1780 to 1795 (Beta-126400) | C14 (Calibrated, 2-sigma) | LA 117165/FB 16318, Juniper Canyon, NM, ocotillo from storage platform, (Baugh and Sechrist 2001) |
| *A.D. 1520 to 1580 (430 to 380 Cal B.P.) AND Cal A.D. 1630 to 1690 (320 to 260 Cal B.P.), Cal 1730 to 1810 (220 to 140 Cal B.P.) AND Cal A.D. 1920 to 1950 (30 to 0 Cal B.P.) (Beta-203378) | C14 (Calibrated, 2-sigma) | AZ CC:12:58[ASM], Peloncillo Mountains, yucca from storage platform in rockshelter with charcoal *gah'e* pictographs |
| A.D. 1430 to 1520 (520 to 430 Cal B.P.) AND Cal A.D. 1580 to 1630 (380-320 Cal B.P.) (Beta-166882) | C14 (Calibrated, 2-sigma) | Dragoon 1, Dragoon Mountains, AZ, burned sugary plant material from small roasting pit |
| A.D. 1354 ±40 (A.D. 1324 to 1404) (Oxford Sample X1874) | OSL | Dragoon 1, Dragoon Mountains, AZ, rock from small roasting pit |
| 1420 ± 80 B.P. (A.D. 1340-1500) (Oxford Sample X1517) | OSL | FB 9609, Hueco Mountains, NM, Llano Plain (likely Apache) sherd |
| Cal A.D. 1270 to 1400 (680 to 550 Cal B.P.) (Beta-169609) | C14 (Calibrated, 2-sigma) | LA 84424, F 1, charcoal from small roasting pit |
| Cal A.D. 1400 to 1450 (550 to 500 Cal B.P.) (Beta-172175) | C14 (Calibrated, 2-sigma) | LA 84424 F2, charcoal from small roasting pit |
| A.D. 1240-1440 (Beta-64068) | C14 (Calibrated, 1-sigma) | LA 70189, Mogollon Mountains, NM (Oakes and Zamora 1999) |

| DATE | TECHNIQUE | SITE |
|---|---|---|
| A.D. 1220-1430 (Beta-69815) | C14 (Calibrated, 1-sigma) | LA 70189, Mogollon Mountains, NM (Oakes and Zamora 1999) |
| A.D. 1400-1470 (Beta-64067) | C14 (Calibrated, 1-sigma) | LA 70188, Mogollon Mountains, NM (Oakes and Zamora 1999) |
| A.D. 1260-1490 (Beta-69811) | C14 (Calibrated, 1-sigma) | LA 70188, Mogollon Mountains, NM (Oakes and Zamora 1999) |

*Included to show continued use of this feature type (storage platform) through time.

Moreover, by the mid to late 1600s, when Apachean presence is accepted (Gregory 1981; Reid 1997b; Wilcox 1981), what we recognize as Apache has already transformed from its earlier more widespread pre-differentiation character. For example, because the material culture of later (nineteenth century) Apachean sites is not directly representative of this earlier pre-differentiation Athapaskan occupation, the historical approach cannot be used directly and without scrutiny in attempting to identify the presence of these earlier groups in this area. I argue that these early Athapaskan-speaking groups practiced a more mobile adaptation than the nineteenth century ethnographically documented Western Apache (and Navajo and Jicarilla) and lived a much different way of life owing, in part, to their intermixture with other groups. The material culture changes that occur at key periods (in this case in the 1600s) seem to represent evidence of this transformation underway. The lack of a clear and direct connection between early and late representations of Athapaskan material culture is equaled by the indirect means by which historically referenced groups are connected to archaeology, on the one hand, and modern ethnic and political groups, on the other.

All of these considerations are taken into account as this chapter incorporates data from a variety of sources to investigate assumptions about and evidence of the arrival of Athapaskan-speaking populations in the southern Southwest. Certain existing perceptions are suggested by the documentary record, others by understanding later Apache land use practices from ethnographic and ethnohistoric sources. Other sources include freshly acquired archaeological data. Combined and critical use of these sources suggest a different scenario than that which has dominated textbooks for several decades.

## Theoretical Considerations

The act of migration for the *Dené* or proto-Apache is not disputed (Cordell 1984; Reid and Whittlesey 1997; Spicer 1962). Ancestral Athapaskan-speaking populations are known to have arrived sometime later than the sedentary agricultural cultures that are considered so representative of the prehistoric Southwest (Anasazi [ancestral Puebloan]), Mogollon, Hohokam, and others). What is disputed is "when" they arrived and what constitutes evidence of them during this early era. More specifically, the question about the Diné migration provides a forum for considering how this process looks for relatively small mobile groups as compared to migration undertaken by sedentary societies. In this way it provides a basis for exposing the sedentary bias behind migration theory and adjusting perspectives toward a mobile way of thinking that is so necessary for finding and understanding evidence of these early mobile groups.

Mobile groups, by virtue of their mobility, migrate differently than groups who are burdened with fixed boundaries, dependent upon stored resources, tethered to arable land and growing seasons, and limited by the requirements of their architectural template. The act of settling-in presupposes a type of relationship with the land and a recognition of and respect for (even as they are violated) territorial boundaries. Migration among mobile groups likely involved a different set of processes than migration for sedentary groups and it therefore is expected to look very different archaeologically because groups are not necessarily burdened by the same presuppositions and prerequisites. As Anthony (1990:895) has noted, "a methodology for examining prehistoric migration must be dependent upon an understanding of the general structure of migration as a patterned human behavior." The important point to consider is that the relevant behavioral patterns differ for mobile groups.

Rather than a migration, per se, which implies a calculated route-oriented move with a designated path or familiar route, and targeted on a known destination (Anthony 1990:895, 896; Haggerstrand 1967:168; Lightfoot 1978:188; Morrill 1974:167), migration by highly mobile groups might best be viewed as a gradual or incremental drift across the landscape. This assumes a directional trend, but without a bearing, other than perhaps orientation with reference to key landforms, terrain types, resource elements, or the attraction of some element of the night's sky-scape. Participants may not have been going anywhere in particular, but they were going as a way of interfacing with the landscape, of accepting the flow of their world, and of utilizing selected aspects of its resources. This waft may have been no more of a plan than moving from playa to playa or following the elk in their seasonal migration. Thus, Anthony's (1990:902) sedentary bias becomes apparent when he states that "Migration across an ecological or cultural boundary would require considerable planning on the part of the migrants, and should leave the clearest archaeological evidence." Yet, movement and fluidity were a way of life for the ancestral Apache, carried out within the context of daily improvisation. Though anticipating the medicine man's suggestion as to when they had finally arrived at the end of their travels (Stockell 2007), such decisions had a pragmatic basis. This is in contrast to sedentary groups who plan for long-distance moves, which occur in the context of stockpiling and specialized or focused adaptations that limit the range of resource zones and habitation sites a group may select. In contrast, when traveling with the migration patterns of your food (e.g., migratory species of large game) and when food sources are interchangeable (reindeer versus elk, deer, or bison) long-distance relocation is no longer "dependent on the long-distance transmission of information concerning potential destinations," as Anthony (1990:902) would have us believe. When a group is on the move as a way of life there are no issues of the "frictional effects of distance" (Anthony 1990:902) because once the day ends mobile groups are where they are supposed to be; they do not have to return to a home base or continue on to a predetermined destination.

Movement was not specifically directional with the intent of getting to a specific location, such as the confluence of two rivers, but was rather like a fog descending upon the basin, filling the path of least resistance. This meander involved exploitation of the terrain, gradual expansion beyond the bounds of an existing range of use, not adhering to territorial boundaries, not vesting in any one set of landscape features, and not observing the territorial rights of others. This is not to say they wandered randomly or aimlessly over the landscape, but rather focus was on use of the landscape and specific (but interchangeable) terrain features rather than on boundaries, though the path of least resistance might result in their use of peripheral areas or hinterlands. In this conceptualization multiple concurrent "ways" of travel and dispersal by different subgroups should be considered instead of a single route. A route or corridor implies a directional path or a passageway, exclusive of others; this does not seem appropriate for the case under discussion.

In attempting to date the arrival of the ancestral Apache, it is important to understand that it is not an *event* that is being dated but a *process,* and that it is a different kind of process than among sedentary groups. Sedentary groups originate in and are drawn to places on the basis of recognized territories that have arable land, as opposed to the worldview or landscape perceptions of many types of mobile groups that are extra-territorial or are unmindful of those types of boundaries. Rather, the *process* of Athapaskan migration probably occurred over an extended period of time by small groups who were potentially practicing a modified version of their mobile lifeway, outside of the sphere of influence and often beyond the political and social sway of resident sedentary groups.

This case study of Apachean migration also provides a diachronic basis for considering what the early end of a migration might look like in comparison to migrations in their more advanced stages, before initial intergroup acknowledgement, contact, and assessments have occurred and before landscape and terrain selection issues have been resolved. There is a tendency to look for and at the most tangible and obvious evidence. For example, it is Anthony's (1990:902) view that "long-distance migration should result in changes that would have distinct effects on the archaeological record." But evidence left by mobile groups is often less visible on the whole than the most subtle evidence left by sedentary groups and so the distinct differences of the incursion are easily lost owing to their unobtrusive nature. Moreover, the signature of a new mobile group presence gains robusticity as time elapses and population levels increase. Such lucid evidence usually does not present itself until the process itself has been underway for some time. In order to understand the "how," "why," and "from where" questions it is necessary to identify the early end of the process as it is just beginning to manifest. This means that observations must precede the time when the robustness of the data make the event most noticeable.

As Anthony (1990:903) correctly notes "initial migrants (the scouts) might have had motives and organization very different from those of the group that followed." Though this is also expected to be true of mobile groups, different factors were likely involved. It is reasonable to infer that these earliest entrants might have distributed themselves differently on the landscape, focusing perhaps on elements that reminded them of their place of origin or that could be exploited in familiar ways. As I have noted elsewhere (Seymour 2003a), it takes time to permeate a new landscape and to discover and fully utilize relevant terrain features and resources.

Also, their material culture may represent no more than a subset of that used later when entire social groups were present and therefore the full range of behavior was being practiced. This may explain the predominance of the hunting and butchering assemblage in the earliest *Dené* sites, but recognition factors are also likely operative given the distinctiveness of this tool kit.

Most dateable early contexts will derive from and after the period at which the migrants reached some critical population level. Thus, it should not be surprising that dates from my sample cluster around the 1400s, increasing from nine for the 1200s or 1300s to the 1400s in the current sample and burgeoning after this (Seymour 2002, 2003b, 2005). This is an interval that likely marks a first noticeable florescence of early occupation rather than an initial entrance. The beginning of this migration process will not likely be seen if the common practice is adopted of averaging chronometric dates both from individual sites (which often were occupied intermittently over extended periods) and also between sites throughout a region (Seymour 2005). By averaging dates, one is seeking to understand a stable cultural system, obtain middle-of-the-road answers, capture commonalities in the confidence intervals, and identify well-entrenched behavior that is fully adapted and completely adjusted to new conditions. Whereas in reality it is necessary to seek the point of change before stasis in order to understand the basis for change and to isolate the initiation of this process. This will be represented by a modicum of difference, a barely perceivable blip, and relatively few chronometric dates. This means it is desirable to isolate the rarely occurring earlier dates (and sites) and to understand that by their rarity they are suggestive of the earlier end of the process prior to saturation and adjustment. Confidence in interpretations surrounding these early dates is increased by their repeated, though low frequency, occurrence across numerous contexts throughout a broad geographic area, always in association with distinctive material culture. A battleship-curve-like frequency to dates (and sites) is expected, which is narrow at the earliest end, and broadens gradually as populations and therefore datable contexts increase, and then narrows as populations decrease and topographic settings in which datable contexts are present narrow at the end of Apachean reign over southeast Arizona. For example, of the seventy or so dates I have run or analyzed from this period only nine predate the Coronado Expedition.

By 1400, when dateable *Dené* deposits become increasingly common, it is likely that this intrusive group had become established and it had reached an initial pinnacle of occupancy and a relatively widespread distribution. As people migrate into new areas it is common to have a lull in population expansion after the initial influx. If conditions are favorable, a gradual and marked increase in population will occur through time as niches are filled in and a wider range and greater depth of the environment is exploited (see Pennington 2001). In some instances, this may be indicated by a surge in sites and an expansion into previously uninhabited areas.

Aside from this issue of increasing visibility through time owing to greater numbers and wider distributions of people, recognition remains a problem because material and spatial patterns were modified as adaptations changed. Identifying and dating the initial entrance is made more difficult by the probability that the material culture indicative of the groups likely changed throughout this early period as groups became more sedentary, adopted new technologies, entered new environmental zones, and intermixed with other groups. This means that evidence of the earliest Southwestern occupation may look very different archaeologically than that associated with the distinctive historically referenced Athapaskan groups. My current notions are that a fairly widespread pre-differentiation Southwestern Athapaskan manifestation (Cerro Rojo complex; Seymour 2002, 2003a, 2003b, 2004 b) looks very different from their later counterparts, as new adaptations, technologies, identities, and social practices are adopted through time. Consequently, at various times in the 1500s and 1600s subgroups changed in ways that made them begin to look more like the corresponding modern groups, a transformation that makes them more recognizable to us, but also inhibits recognition of earlier representations.

As people move into an area, some occupy separate niches, remaining detached from resident populations and maintaining their distinctiveness (Barth 1969b; Seymour 2002, 2007a, 2008b; Woodson 1999). In other instances they integrate into an existing populous, and in doing so the respective social groups may be gradually transformed. Both processes are observable in the archaeological and historic records relating to early Athapaskans (Seymour 2002, 2004a). As migrating groups move into a host community and adopt a new lifeway not all elements of their material, spatial, and organizational systems are affected simultaneously (Seymour 2002, 2007a, 2007b, 2008b). Not only do the material and spatial aspects of human behavior change at different rates but so too do tangible changes occur at a different rate than the rate at which people form new social groups. This disjuncture means that the act of people moving often precedes or succeeds evidence of change in the substantive spatial and material realms within a receiving community. Depending upon the form and degree of integration into a host community, certain technological systems of one or both groups may persist as distinctive (Seymour 2002, 2008c; Woodson 1999). In my data set these persist for 120 and 260 years before certain of the technological attributes related to the formerly unique adaptation of the migrating group are lost. Use life, curation, and recycling factor into persistence, but retention of technological knowledge and continuation of a way of making tools indicates adherence to a traditional manufacturing procedure that may be identity related. The replacement process can be seen as an integration index in the El Paso area where the material culture of the non-Athapaskan mobile groups who intermixed with and co-inhabited settlements with Athapaskans (according to the documentary record as well) is retained and is visibly distinct in the Athapaskan assemblage for about seven generations or 220 years before it is lost (Seymour 2002). Numerous examples are available from a variety of cultural contexts to suggest that site structure changes at different rates when accommodating new groups, depending upon the basis for the change (e.g., integration of other groups versus responding to needs for defense; Seymour 2002, 2007a, 2008b).

Nor is it reasonable to expect that earliest Apachean groups would have migrated southward in a straight line, thereby leaving evidence of their presence in intervening areas along this imaginary baseline. This may account for Dismal River, if one accepts this as Apachean (Gunnerson 1960). This straight-line effect is not realistic for two reasons. First, as Anthony (1990:897) notes movement often occurs in two directions, between the point of origin and destination. More pertinent to mobile groups, a clear trail of entry is not expected because most prehistoric trails meander across the landscape, rather than follow a linear trajectory, and they are not a path, but rather they represent a general way. They behave this way to avoid certain terrain features but more so because trails connect key settlements and resource areas that are not necessarily accessed via a direct route and they link prominent and perceptible landforms and water sources. Trails also circumvent dangerous areas, such as those occupied by known or potential enemies. The reflexive nature of this process leads to a backtracking effect, which is even apparent in the travels of Alvar Nuñez Cabeza de Vaca in the 1530s as he found his way back across the continent toward New Spain (Hodge 1990), in contrast to some of the early expeditions that followed the most direct routes across the lowlands, through passes, and along rivers. The meandering and often backtracking nature of trails and population movements may very well explain some of the apparent anomalous distributions of Athapaskan material culture in the eighteenth century and earlier, as migrant groups perpetually ebbed and flowed across the landscape.

Mobile groups in the southern Southwest tended to use considerable areas of the landscape—larger than most of us are apt to conceptualize (Seymour 2002). The fuzzy edge effect of unclear territorial boundaries in such circumstances is conducive to landscape drift. Gradual movement across the terrain for initially short distances might infringe on others' boundaries in ways that were not clearly noticeable, especially when groups used an area and then moved on, partially retracing their steps. Flexible social boundaries—so typical of mobile groups—would also contribute to imprecise social and geographic boundaries, particularly when intermarriage, adoption, kidnapping, and other forms of recruitment sufficiently altered the composition of the social group that members were at home in any of a variety of adjacent home ranges.

The desire to stay out of the way of hostile neighbors would also lead to a stretching of nodes of use beyond existing physical and political boundaries. An example of this is the historic period Warm Springs Apache who seasonally traveled all the way to the tropical belt in Mexico to escape the pressure of the American military (Ball 1970). Such stretching or protraction can occur during periods of resource shortages in one area, or abundance in another. The familiar node or end point can become the springboard for further movement. An expansionist worldview (see below; Cribb 1991:22) would contribute to information gathering about the peripheries and beyond. This conceptualization of elasticity in terrain selection and landscape posits an advance across the geographic expanses in a leisurely leapfrog fashion (Anthony 1990:902) for logistical mobility and a surging (advancing and partially retreating) progression for residential mobility. This is not simply a spatial expression of a focused subsistence pattern (such as is practiced by sedentary agriculturalists), as Anthony (1990:903) claims, but rather seemingly represents a pattern expected when a group goes from the familiar to the less familiar. The distance of the leap or the breadth of the surge are conditioned by knowledge (and how knowledge is obtained) about what lies ahead and behind. But unlike among settled groups, a surge is often the product of an accumulation of information gained during daily rounds, where a misstep has fewer dire consequences than a leap. In addition to scouts (as hunters or specialized task groups or age sets) inspecting new areas, entire social groups (including gatherers and child rearers) may participate in the process, thereby bypassing previous base camps and establishing new utilization zones in areas beyond the former periphery. For example, certain Athapaskan groups, such as the ancestral Mescalero, used both the plains and mountains (see Seymour 2002), and there is certainly no reason to believe that they did not use both zones as they found their way south. It is also a reasonable scenario that several small groups dispersed and coalesced seasonally, as is typical of hunter-gatherers globally, and then dispersed and ultimately never came together again. Factional disputes and a variety of other reasons including simply becoming lost would contribute to such a splintering.

In this form of mobile migration, movement is not necessarily seasonally driven, but is dictated in part by their zone of use overlapping with the territories of several other groups, whose interests in boundary maintenance or use rights could lead to conflict. Incursions might be incidental to resource procurement but could be deliberate, as in raiding. The late historic Apache are known as raiders and it is understood that this was subsistence-focused raiding rather than warfare (Goodwin and Basso 1971). Another factor is that raiding would lead to reprisals and to the development of defensive measures by sedentary groups that, if successful, would be as devastating as an extended drought experienced by sedentary groups. Contrary to current conceptualizations, I see evidence from the Cabeza de Vaca account that raiding was already a widely practiced adaptive strategy in the southern Plains and hill country and the Southwest by A.D. 1535-1536 (see Seymour 2002). By this time raiding and inter-group conflict had already prompted a set of customary responses among the resident populations that attest to a greater time depth (see Seymour 2002). The presence of these raiders in the southern Southwest by at least as early as A.D. 1535-1536 brings to light several important propositions. To have reached the international border or beyond by this time from the north the *Dené* would have had to have a much earlier start than is currently conceptualized. Local residents were already responding in a variety of formalized ways to the threat of raids, which leads to the conclusion that either the *Dené* were present far to the south much earlier than the A.D. 1525 date the Gunnersons have suggested (Gunnerson 1956; Gunnerson and Gunnerson 1971:7) or non-Athapaskan groups had this raiding adaptation. It may be that both of these possibilities occurred, as my data suggest: certain groups of ancestral Apache were present as far south as the international border long before A.D. 1525 and groups of non-Athapaskan raiding nomads were present as well (Seymour 2008e).

Trespass into the established territories of others, whether to escape threat and deprivation, or intrusion into others' settlements for the purpose of plundering, would undoubtedly lead to displacement. In these ways, the *Dené* pattern is similar to that for other types of nomads where there may be considerable shifts in areas exploited or as Cribb (1991:22) phrases it, there may be "marked instability and dramatic shifts in migration tracks," and as he notes, the nomadic system is "inherently expansionist."

When one considers that these early *Dené* may have had an adaptive pattern more akin to (nonpastoral) nomadic groups than mobile hunter-gatherers this model become clearer. The motivations and patterns of mobility are important for understanding this difference and for recognizing the archaeological pattern, though any of a number of typological distinctions can be made (see Barnard 1983; Cribb 1991). The reasons for mobility differ from standard hunter-gatherers, both because raiding was a key part of the adaptation and because larger expanses of landscapes were used. Residential stability was dependent upon the success and responses of raiding targets as was residential site selection. Nomads can be hunter-gatherers or pastoralists but the defining criterion is the lack of a fixed home base over the long run that keeps nomads moving from place to place. Use of serially occupied home bases and the accrual of new bases dependent upon the practices of neighboring groups, falls within this classification as well if territorial attachment is not strong.

Following Khazanov (1984), Cribb (1991:20) notes key differences in the basis for movement, the patterns of movement, and in the way sites are formed between nomads (pastoralists in Cribb's examples) and hunter-gatherers. One relevant difference, as Ingold (1987) has pointed out, is that hunter-gatherers focus on territories or significant places whereas nomadic "productive relations" are largely location independent, often focused on their "mobile pastoral capital." Nomadic pastoralists focus on access to pasture rather than "establishing vital symbolic links between the community and a particular locale" (Cribb 1991:21), as do hunter-gatherers. Most hunter-gatherers are intimately familiar with the entire range of resources in their territory. In contrast, by focusing on a narrower range of resources over a broader area, relying on interchangeable species of large game (reindeer versus elk, deer, or bison) and supplementing this income with the proceeds of raiding, the early *Dené* would have occupied a different niche than other mobile groups that would be more typical of hunter-gatherers in general. Their settlement pattern and site structure would differ as well. Cribb (1991:22) refers to the expansive nature of landscape use by nomadic groups as "territorial aggrandizement." As time went on and population pressures increased, a shift would be expected in use and conceptualization of the landscape.

The abundance and ubiquity of large game would result in a subsistence strategy that generated patterns more typical of nomads than hunter-gatherers. Bison (documented even in the Southwest), deer, elk, or antelope were found throughout a much broader range than any one hunter-gatherer territory, contributing to a diffuse or broad-spectrum economic strategy rather than a focused one (cf. Anthony 1990; Cleland 1976). Yet, rather than encouraging diversification in any one location and therefore residential stability within a region, as is often suggested (see Anthony 1990:901), this selection of widely available and interchangeable resources facilitated and perhaps even encouraged movement across large geographic expanses, beyond familiar settings. This is particularly the case when coupled with an expansionist or aggrandizing notion of landscape use. Moreover, when raiding could offset periods of food shortages, precise knowledge of specific resources in a particular landscape was not as critical. Flexibility in landscape choice was aided by a focus on the commonalities of a larger geographic area, rather than becoming dependent upon the specifics of a local district. At the same time a more generalized use of the landscape that is adapted to the commonalities of adjoining areas facilitates elasticity, encourages leapfrogging and surging, and expands the options of landscapes available for use. This not only permits the use of alternative landscapes in a serial fashion but it allows greater tolerance for environmental fluctuations of certain types. Localized variation becomes less important, while regional or pan regional vacillations can be devastating under such circumstances.

Though migration is often treated as if it were a rarely occurring event, in reality what is rare is our ability as archaeologists to recognize and isolate the multifaceted signature of such processes. One reason for this failure is use of a too-narrow focus on what are really multi-cultural and multi-regional problems. Archaeological culture groups are a mainstay of the discipline and it is widely understood that from the culture group concept arises the assumption of geographically stable and sedentary social groups. Even when this culture group concept is rejected the notion of stability persists. All but the most robust and obvious processes of migration and intermixing in the Southwest will be missed because observations are focused too locally, when what is needed is a more continental perspective of these processes.

Only recently has migration regained favor as an agent of cultural change, as some of the papers assembled in this volume suggest, building on the earlier work of Di Peso (1958), Haury (1958), and others. Yet, migration studies still tend to focus only on sedentary groups, their aggregation and dispersion, and their formal and predictable transfer to new regions. It is time to reconsider the importance and role of mobile groups in the migration process and also in the distributions of sedentary populations. Mobile groups were an important feature of the social and economic landscape in the late prehistoric period and their mode of migration differs substantially from sedentary groups and therefore looks different. If we are to see them we must look for a different type of evidence in different ways. If we are to understand the impetus for region-wide changes we must consider the role of mobile groups in stability, disruption, and reorganization.

## Archaeological Summary

To address this issue of the *despoblado* and Athapaskan migrations further it is useful to turn to a summary of the archaeological data from both the heart and fringes of the mountains. Two key archaeological issues will be addressed with reference to the idea of an uninhabited zone: (1) the basis on which it is inferred that Athapaskans were present this far south and west by 1540 and (2) that the Coronado *despoblado* was in their range at this time as indicated by archeological site distributions in and near the Mogollon uplands around and north and east of Safford. The issue of how it has been determined that these are Athapaskan sites have been dealt with at considerable length in other fora (Seymour 2002, 2003a, 2003b, 2004a, 2005, 2007d; Seymour and Robertson 2008).

Fundamental to answering these questions are (1) the correlation between archaeological material culture, on the one hand, and historically referenced groups, ethnographically described groups, and modern ethnic groups, on the other (Seymour 2002, 2003a, 2008e), and (2) extension of these data backward in time, though never assuming there is a direct connection. This involves additional methodological considerations that include (1) modeling mobile group behavior with reference to landscape use, material culture innovations, and site types (Seymour 1995, 2002), a portion of which has been summarized in a previous section of this paper and (2) considerations regarding discovery, selection, and interpretation of samples for chronometric dating in relation to wood use behavior and landscape use (Seymour 2002, 2003b, 2005). These topics are complex and require detailed analysis and consideration, and while they will be touched upon here, they have been addressed in depth elsewhere and the reader is referred to those other sources. The dates presented in Table Six are for illustration purposes only. It is imperative that they are placed in behavioral and archaeological context; for this readers are referred to other treatments (Seymour 2002, 2003b, 2004a, 2005, 2007d, 2008d, 2009).

Throughout southern Arizona and southern New Mexico a number of early (pre-Coronado) Athapaskan habitation sites have been identified including small camps and large multi-group rancherías (Figure Thirty Eight).[1] Two especially large sites are known from the Safford area—those in the Whitlock (AZ CC:7:11[BLM]) and Peloncillo (AZ CC:12:58[ASM]) mountains—although many more are likely present but as yet undiscovered. These sites occur in and near the area of this historically referenced *despoblado*, so that the residents were within easy reach of any route Coronado might have taken (Seymour 2008a). These large Athapaskan sites sprawl across the mountaintops and slopes, with hundreds of structures arrayed on the saddles and tucked into the interspaces between boulder outcrops, with rockshelters and shallow overhangs providing cover for other features and rock art. Additional smaller sites known from the Mogollon uplands can also be definitively attributed to the early Athapaskans and several others are known further south, west, and east of this zone (Table Six) (Seymour 2002, 2003b, 2004a, 2005).

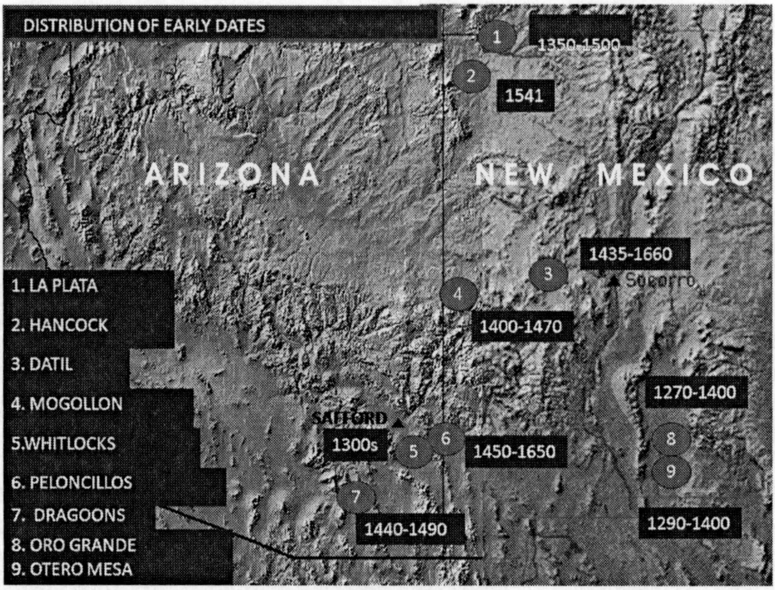


Figure 38: Sites 1540 or earlier in the Mogollon Highlands and south.

Many of these sites (or components on them) date to or before the time of Coronado indicating that there was a widely dispersed early *Dené* population throughout the southern Southwest by the time of the Coronado Expedition (Table Six). Dates that are associated with distinctly Athapaskan material culture and conform to predictable landscape-use patterns span from the late prehistoric into the historic period (Seymour 2002, 2003b, 2005, 2008a). Only those predating Coronado are presented in Table Six. Radiocarbon dates on annual species from distinctive feature types (storage platforms) indicate dates as early as the 1200s or 1300s, with a continued use of this unique feature type on other sites well into the post-contact historic period. Other dates from features found in association with distinctive assortments of flaked stone and features (roasting pits and structures) also hint at this early presence, whereas many more samples are from the 1400s and later as these people become more visible in the record. Compilation of dates continues as part of a systematically designed program to date the early Athapaskan presence (see Seymour 2002, 2003b, 2004a, 2005).

Each of these sites, including those producing early pre-contact dates, possesses a subset of a constellation of shared traits that indicate they represent early *Dené* occupations. The flaked-stone assemblage is the most useful artifact class for identification purposes because it is most widespread, and is found on most sites, whereas pottery is highly restricted in distribution owing to its limited context of use. Flaked-stone artifacts were used in a wider range of tasks than most other forms of non-perishable material culture and therefore are available to be found and represent a wide range of tasks undertaken. Still a laundry list of traits is not especially helpful for conveying the nature of the flaked-stone assemblage because, as with most culture groups, it is the stylistic characteristics of the artifacts that are distinctive, rather than the generalized functional forms that include scrapers, knifes, gravers, perforators, burins, projectile points, and more. The flaked-stone tools are indicative of a hunting assemblage used by raiders who were especially thrifty and pragmatic (Borden 1952; Gunnerson 1960; Seymour 2002, 2003a, 2004a, 2004b). As Sahlins (1972:12-13) notes, rather insensitively, without a sense of possession some hunters, at least, display a notable tendency to be sloppy about their possessions. They have the kind of nonchalance that would be appropriate to a people who have mastered the problems of production. "They do not . . . take care of their belongings," nor do they put them in order. They

> place no value whatever on their utensils and . . . they have completely forgotten the effort it took to make them. Actually, no one clings to his few goods and chattels which, as it is, are often and easily lost, but just as easily replaced . . . The Indian does not even exercise care when he could conveniently do so.

It is in this sense that rather than investing considerable effort in the shaping of fine durable tools, as is characteristic of many other mobile groups, the early *Dené* focused on replaceablity and substitution. This might be expected in a context where property was procured through raiding, retaliation resulted in routine loss of sizable fractions of the household equipment, and where sudden movement was required between alternate encampments. The "drop everything and move out" focus of the assemblage is apparent from the earliest times. Consistent with their degree of mobility, tools, like structures and other feature types, were improvised on the spot using locally available materials and incorporating naturally occurring attributes of the terrain. Investment in tool production and hence curation of tools is seen only with a few forms that require special materials (e.g., fine-grained material) that increase flaking quality, durability, and reliability. These include some of the small thin side-notched projectile points (often complemented by small triangular indented base forms), the bifacially flaked knives (in contrast to the flake knives and agave knives), drills but not perforators or gravers, and hafted side and end scrapers that are distinct from those flake scrapers that exhibit no real modification. Groundstone is consistently expedient and includes handstones and slab metates on small flat rocks and grinding slicks, reuse of prehistoric forms, and perhaps use of bedrock mortars. Further elaboration of these forms is not appropriate in this context as it is important to consider the assemblage as a whole, as has been done at length elsewhere (Seymour 2002, 2003a, 2004b).

One of the most effective ways to connect modern ethnic groups to the pre-European archaeological record is when traits fundamental in a belief or organizational system and unique to a group persist through the ages. Persistence in an unaltered form is rare and is generally not expected. *Gah'e* (*ga'an* among the Western Apache) or representations of mountain spirits are a trait that has been transmitted through time that can be definitively connected to this specific ethnic group. Rock art that represents these distinctly Athapaskan mountain spirits is inscribed on the back of various rock shelters; two such rock shelters are known from the Peloncillo Mountains at AZ CC:12:58(ASM) (Figure Thirty Nine). Multiple representations of black charcoal and red painted *gah'e* adorn these shelters that also contain distinctly Athapaskan features (platforms), structural features, and artifacts (pottery and flaked stone). The association on this site of these distinctively Athapaskan material culture representations (such as the *gah'e*) with the distinctive feature type (the storage platform; not found in prehistoric ancestral Puebloan [Anasazi] or Mogollon contexts) suggests that these are related and that at this site they date to the 1400-1600 period (see Table Six; Seymour 2002, 2003b, 2004a, 2005). This notion is reinforced by the post-contact dates obtained from some of these features on sites further east, where they have produced dates late into the historic period (Seymour 2003b, 2005, 2006b). In other contexts, such as in the Whitlock Mountains at AZ CC:7:11(BLM), where there are also many mobile group house outlines, one of these storage platforms has been dated to the 1200s or 1300s. Thus, this distinctive feature type (the storage platform) has been dated to the late prehistoric period and to the late historic period. No other group occupied this region for this duration and this particular date span, and no other group (prehistoric or historic) occupied all of these widely dispersed prehistoric culture areas and so this provides support for the association between these unique features and the proto-Apache, especially since dates are from the features themselves.

Figure 39: A black charcoal *gah'e* on wall of rockshelter that also contains a storage platform, AZ CC:12:58[ASM].

The dwellings found on these residential sites conform to mobile group structure types found throughout the West, being surficial, circular or oblong, and admittedly difficult to recognize (Figure Forty). Rocks were placed around the perimeter to hold poles or branches in place, or appropriate naturally occurring bedrock outcrops or boulder distributions were supplemented with the addition of a few rocks. In other regions, such as in California, these types of modest, inconspicuous rock rings and clearings in rocky surfaces are commonplace, known from decades of research as distinct feature types (Rogers 1939, 1966). Similar unassuming structural features have taken a while to be acknowledged in the Southwest, where more unmistakable house types that represent more sedentary populations are widespread. These less obvious types of dwellings or houses are indicative of use by mobile groups who did not invest a great deal of resources or effort in their construction. Their improvised nature relates in part to their limited duration of use and also to the restricted purposes to which they were put, unlike Pueblo dwellings that served a more general range of habitation functions (Seymour 2007d, 2009).

Figure 40: Mobile group structures from around Safford.

Plain pottery exhibiting a range of attributes is present on some of these sites. Specialists in the local prehistoric material culture, such as Patricia Gilman, Stephanie Whittlesey, and David Hill, agree that this pottery is not prehistoric or Mogollon, and recognized specialists in Athapaskan ceramics agree that this pottery is Athapaskan. Thin-section petrography and luminescence dating also distinguish this pottery from that made by prehistoric sedentary groups (Seymour 2002, 2003b, 2004a, 2007b, 2008d). It is especially interesting that the diagnostic pottery most consistently associated with the Apache seems to appear only relatively late in the sequence. The classic Athapaskan pottery, referred to as Apache Plain and found among the Western Apache, is gray-colored, thin, with a wiped surface. Luminescence (optically stimulated luminescence [OSL]) dates on this most widely recognized Apache pottery and its southern brownware counterpart so far suggest that it does not occur until the 1600s in the Mogollon highlands and elsewhere, which perhaps not surprisingly matches the conventional belief that the Apache did not arrive in the Southwest until this time. However, if Apache Plain and similar late varieties mark not the arrival of the Apache, but an innovation in or borrowing of ceramic technology or style, a reliance on this diagnostic artifact as an ethnic indicator would obscure other evidence for an earlier arrival of the proto-Apache.

Plainwares found on the earliest Athapaskan sites, including those in the Mogollon uplands, are thus far very different from these later Apache plainwares, and certainly more diverse. Variations in vessel forming techniques and composition indicate that many of these pots were obtained from other groups through raiding, trading, and scavenging (and that learning networks varied substantially). Even so, their presence on early Athapaskan sites allows sites to be dated using OSL, and the appearance of ceramics of a unique character provides an indication of a new presence in the area that speaks of different patterns of interaction than are apparent in the earlier prehistoric period (see Seymour 2003b and 2008d for a discussion of this). The widespread distribution and standardized nature of some of the later Apache pottery indicates that a fundamental change occurred in the post-contact period. Importantly, if this late-dating pattern continues to hold as additional samples are run, adherence to this index artifact as an indication of Apache presence will likely continue to result only in the discovery of late-dating Apache sites.

The distribution of archaeological sites with these uniquely Athapaskan characteristics suggests that Coronado did not wander through an empty wilderness, but instead narrowly avoided encounter by sticking as much as possible to the low-lying grasslands and river margins and stopping at well-known waterholes. On-the-ground data indicate that the Safford area and the surrounding mountain ranges were important within a province of early Athapaskan occupancy. This archaeological evidence is beginning to suggest that this area represented the core of dense populations of early Athapaskans. The Gila- and Safford-area mountains as well as the zones to the north and south were remote from the sphere of Spanish influence, rugged, thick with vegetative cover, abundant in game, water, pasturage, and fuel wood, ideal for hunter-gatherers and raiders, and not conducive to notice by outsiders.

In sum, in whatever way one interprets the route of the earliest European explorers this apparent *despoblado* near and around Safford (between the San Pedro and Cibola) was not devoid of population. Multitudes of Athapaskan and non-Athapaskan nomadic occupants inhabited any of the zones through which or near where these early explorers could have traveled. Such groups were occasionally noted at the fringe of Spanish travel corridors. These Athapaskan and non-Athapaskan mobile groups were occupying the mountains and river basins, including those in the Safford vicinity, for a hundred years or more before Coronado.

## Concluding Statement about an Early Athapaskan Presence

Coronado's *despoblado* is materializing as historical fiction. Like the proverbial fly on the wall, archaeological data divulge secrets about the past that would otherwise have remained silent. The numerous Safford area sites, the dates obtained from them, and the well-established settlement pattern early on all suggest a pre-Spanish occupation (Seymour 2004b). Elsewhere, others and I have argued that the ancestral Apache were present, but not noticed by the earliest Spaniards because they lived in the mountains, whereas the Spaniards tended to move through the plains and river valleys (Danson 1957:116; Harrington 1940:510; Seymour 2002, 2004a:186, 2004b). This seems all the more clear now that early Athapaskan sites are beginning to be discovered in the ranges surrounding Safford, Duncan, and Bowie, Arizona and Reserve, New Mexico and now that segments of the Coronado route have been identified. This scenario is consistent with archeological evidence of the mountain-oriented distributions of Athapaskan settlements. Patterns of Athapaskan landscape use throughout this region conform to a consistent set of terrain characteristics. These are fundamental to the adaptation but because they are shared by mobile groups in general, they have implications for landscape theory, ritual imprinting on the landscape, as well as issues of intergroup interaction, recognizing migration, and finding Coronado's route.

This early Athapaskan presence is consistent with certain Apachean legends that were recorded during earlier ethnographic periods, when political and economic advantages did not weigh so heavily in consideration of these materials. One traditional story notes that the ancestral Chiricahua were present at Spanish contact but remained hidden from the Spaniards for the life of one man (Cole 1988). This same source (Cole 1988) refers to the ancestral Chiricahua movement southward, along the forested flanks of the Rocky Mountains, "between the treeless heights and the treeless plains." This route would explain the early dates in the Mogollon, Datil, Peloncillo, and Dragoon mountains and in the Four Corners area. Traditional stories of other contemporaneous groups, such as the O'odham who resided in the Sonoran Desert to the west, refer to armies emerging from the east "coming through like ants from their hills" (Russell 1975:226). From this it is easy to imagine early Athapaskans coming forth from the large hilltop settlements as are now known in the Safford area and using the magic of their weapon system and new conflict tactics to topple the adobe-walled compounds of the basins.

Early Athapaskan presence this far south in the Safford Basin and its margins has implications for population movements of Puebloan groups in certain sectors of the Mogollon region. Those examining Mogollon and ancestral Puebloan (Anasazi) interaction and settlement shifts would do well to reconsider that those people were not operating in blissful isolation. The full-scale abandonment of sedentary people from different parts of northeastern Arizona "joining together at a new site in the upland periphery of a new region" near Safford (Woodson 1999:78) is consistent with expectations of refugees fleeing disruption of a northern homeland, a process that did not escape the notice of early twentieth century scholars. Arguments for the migration of prehistoric sedentary culture groups into the Safford area rely on dates that are contemporaneous with those from nearby Athapaskan sites. Scholars who have limited their discussion of migration to sedentary Puebloan populations would benefit from a broadened perspective. To understand this complex topic more fully it is imperative to consider the effect of the simultaneously occurring migration of mobile Athapaskan (and non-Athapaskan) groups and its effect on the expression of identity in indigenous and migrant sectors of the Safford area populations, a topic that is not even considered in some of the more recent treatments of this topic. In fact, interpretations such as the apparent depletion of large game in the Pueblo IV period (Clark 1998) are potentially attributable less to depletion of faunal populations than to the inability of sedentary residents to venture into hunting areas occupied by mobile groups. Thus, the Safford Basin and surrounding areas are excellent locations to study migrations of mobile groups as well as of sedentary agriculturalists and the roll of mobile groups in the advance of vanguards of agriculturalists as they moved south.

## Notes

---

[1] These include sites found and dated by the author as well as sites found by others that she has inspected and analyzed. The work has involved survey of areas suitable for Athapaskan occupation across a broad geographic area [Arizona, New Mexico and south and west Texas] and has included work in each of the currently recognized historical homelands of each of the key Southwest Apachean groups (Jicarilla, Lipan, Chiricahua, Mescalero, and Western Apache). This has allowed comparisons of Athapaskan and non-Athapaskan material culture in the field and in collections, between geographic and culture areas and through time.

# CHAPTER SIX

# SAFFORD & SAMARQAND

STEPHEN H. LEKSON

Samarqand, in central Asia, was the fictional home of Shah Zaman and the real capital of Tamerlane. Samarqand's name conjures a golden, exotic city in both fable and history. George Bailey, in *It's a Wonderful Life*, wanted to go there. Not many people hope, someday before they die, to see Safford.

Yet Safford has its attractions, and I hope the good people of its valley forgive my jest: it is today a fine place to live, and at various times in the distant past Safford may have shared a bit of Samarqand's glory, as we shall see. Safford today is one of the smaller county seats in the state of Arizona, distant and peripheral to the state capital of Phoenix. So too its archaeology: distant and peripheral from the better known, better studied Hohokam and Mimbres.

## Nobody Wants to Work in a Periphery

The opposite of central, beyond the pale, out on the edges: *periphery* is obnoxious. Scholars who invest money and time and energy into an area's archaeology, want it to be something more than marginal. Alas, Safford was peripheral. Sometimes. So was Chaco Canyon, sometimes. Chaco was central in the eleventh and twelfth centuries, minor in the thirteenth, and nothing but a memory thereafter. The center shifts, and former peripheries rise to glory. Think of England. No more than a nuisance to Caesar, who dismissed its blue-painted natives as savages living on milk and flesh, clad with skins. Many centuries later, the sun never set on Victoria's empire and Rome was a place to see ruins. The trick is to figure out when an area—such as Safford—played which roles in the larger histories of its region.
Some peripheries are well and truly marginal. Think of Fremont. Its short

burst of corn-fed glory was only a brief interruption of millennia of hunting and gathering. In the distant desert districts called Fremont, there were hard limits on what the land allowed. But Safford and the upper Gila River could not have been that kind of periphery. The valleys of the "Upper Gila" were lush, from Cliff through Redrock and Duncan, to Safford and thence into the desert. Framed by mountains, the fertile farmlands should have seen intense occupation, and they did.

Are Safford's peripheral qualities a product of the present, not the past? One could argue that accidents of research focused archaeological attention on Mimbres to the east and Hohokam to the west, and unfairly overlooked the Upper Gila. Yet the earliest archaeologists were aware of Safford. Jesse Walter Fewkes visited Safford in 1897, and placed its archaeology on record before Mesa Verde became a National Park (Fewkes 1904). No one could question that the Safford Valley and the Upper Gila had major archaeological sites: Walter Hough made that clear a hundred years ago (Hough 1907) (Figure Forty One). Yet this stretch of the Gila River has long been considered peripheral; and I think it truly was, in the best and most useful sense. It was on the edges, between larger polities—but that can be an exciting place!

## The Scale of Southwestern Polities

We have come to see the ancient Southwest as home to regional entities far larger than early archaeologists ever imagined. Beglamored by visions of Pueblo "city-states"—democratic, independent, sturdily agrarian and egalitarian—founding fathers such as Edgar Hewett created a vision of the ancient past which appealed to generations demoralized by the first World War, the Great Depression, and the apparent failure of western civilization. Ruth Benedict added to the mythos with *Patterns of Culture*, first published in 1934 and still in print. The idea of sizable political systems—regional polities—was dismissed *a priori* from the ancient Southwest. It took almost half a century to move beyond Hewett's happy, but flawed stereotypes, and to recognize Hohokam ball-court and Chaco great house regions. And even longer to restore the Southwest to its proper context: the remarkable, politically and economically interconnected Post-Classic world (see Michael Smith's and Francis Berdan's 2003 *The Postclassic Mesoamerican World* for a summary of just how interconnected it was).

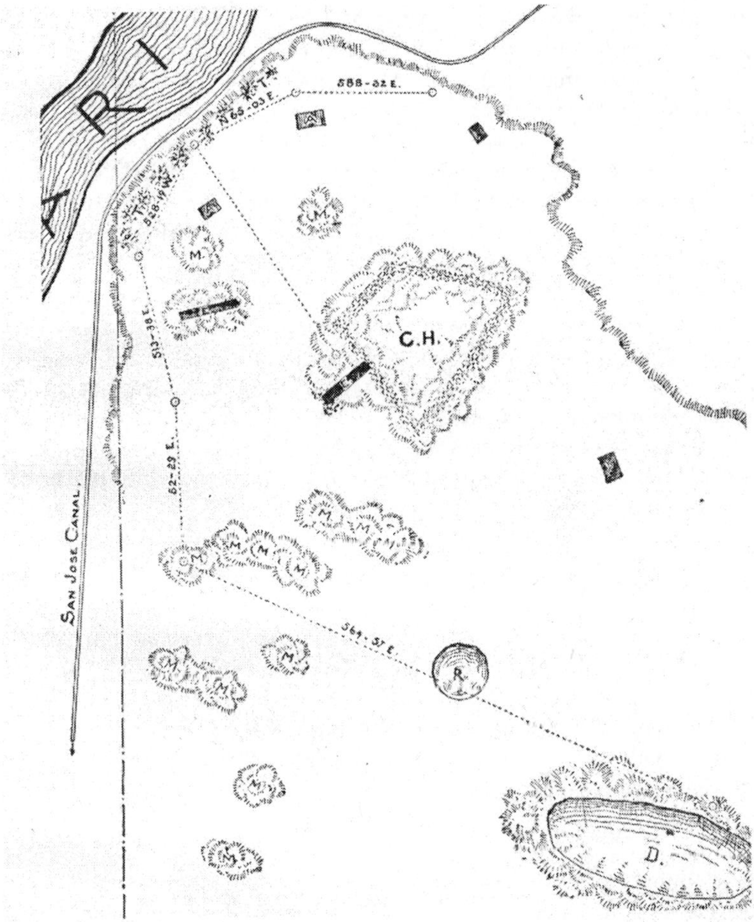


Figure 41: The Buena Vista site (also known as "Solomonsville" and "Pueblo Viejo"). Reproduced from Fewkes (1907: Fig. 3). "D" indicates the ball-court, "C.H." the possible platform mound, now destroyed. I have often wondered if "R" might be a Great Kiva: the depression interpreted as a Great Kiva at the Redrock site was almost 2 m deep.

Regional polities the size and scale of Chaco's great-house region or Hohokam's ball-court system were the rule, not the exception in Post-Classic Mesoamerica and the Mississippi Valley. We should not be shocked to see such things in the Southwest. Three we know well: Hohokam, Chaco, and Casas Grandes (Figure Forty Two). These three regional polities came, as it were, in series: Hohokam peaked in the eighth and ninth centuries; Chaco in the tenth and eleventh; and Casas Grandes in the thirteenth and fourteenth. Each had a clear center (Phoenix, Chaco Canyon and Paquimé), and each operated on similar scales: an inner core of about 140-150 km radius, within which large-scale, tump-line economies—moving bulk goods on peoples' backs—were possible (Malville 2001), and an outer margin of perhaps 240-250 km radius, which marked an apparent limit for political prestige economies (Lekson 2008). Look at Safford and the Upper Gila, at or beyond that evident outer margin of all three—a fact of considerable import even thought Hohokam, Chaco and Casas Grandes were sequential, not (entirely) contemporary. As we shall see, the periphery was not a bad place to be.

## Safford as Periphery: Hohokam and Mimbres

Safford sits between Hohokam and Mimbres. The remarkable, explosive developments of the Hohokam Colonial Period (A.D. 750 to 900) were seminal events in the prehistory of the ancient Southwest; the Colonial Period reverberated to the northernmost fringes of the Pueblo region (the big ninth and tenth century villages of the northern San Juan can be understood as reactions to Hohokam Colonial, but that's another story; Lekson 2008).

Early Mimbres, prior to A.D. 950 was greatly affected by Hohokam (Creel and Anyon 2003; Lekson 1993, 2006). Hohokam objects, ideas, and technologies are found in Mimbres sites and landscapes. We can argue about the significance of Hohokam's influence on Mimbres, but we cannot deny its reality in Late Pithouse and early Mimbres periods. In the Upper Gila, a major Hohokam ball-court (at the Buena Vista or Pueblo Viejo site, Figures Forty One and Forty Two) was only a valley segment away from a huge Mimbres town (at Redrock)—about 80 km as the crow flies. (The Redrock site was three times larger than Galaz or Old Town—if Hough had seen Redrock, Mimbres archaeology might well be different today.)

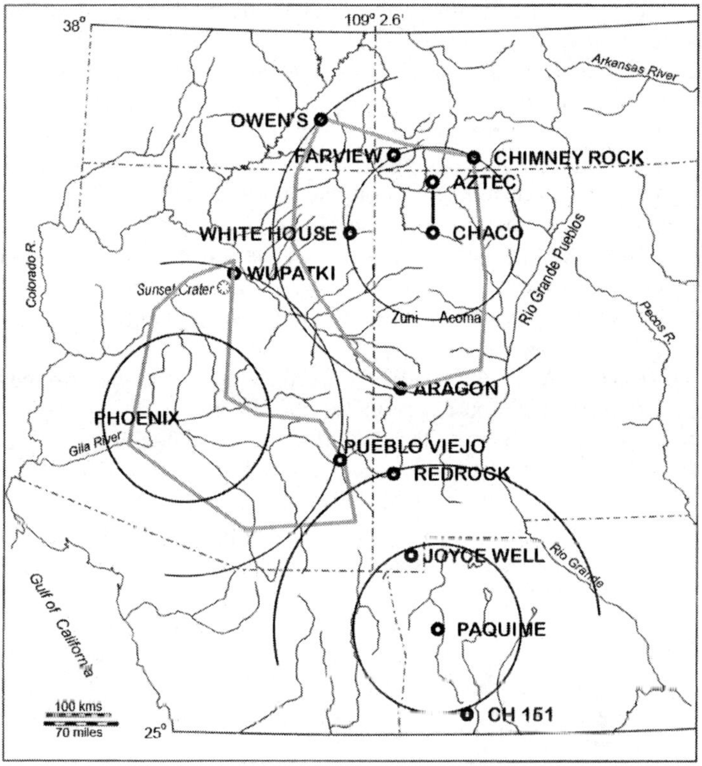


Figure 42: Schematic map of Southwestern regional polities. Note that "Pueblo Viejo" is an alternate name for the Buena Vista site. The inner circles centered on Phoenix, Chaco, and Paquimé are 140-150 km radius (the limits of bulk, tump-line economies; Malville 2001); the outer arcs are 240-250 km radius. The distribution of Chaco Great Houses and Hohokam ball-courts are shown, in gray, and selected examples of platform mounds in the southeastern Hohokam region, as black dots. Casas Grandes's region is yet to be defined, but 150 km radius represents the approximate minimum northern and southern limits of Casas Grandes sites (Joyce Wells and CH 151, respectively; Lekson 2008; based on Whalen and Minnis 2001 and MacWilliams and Kelley 2004), while at 250 km distance, Redrock was clearly economically engaged with Paquimé (Lekson 2002). Thus, it seems likely that the Casas Grandes region was comparable in extent to Chaco and Hohokam, in both its interior and exterior provinces.

Eighty kilometers was nothing: the distance from Chaco to Aztec Ruins. Those two Anasazi sites did not simply “interact,” they were architecturally linked by a monumental “road.” The ancient people were not dissuaded by a paltry 80 km, a distance well within the sphere of bulk-transport, tump-line trade. So, with that proximity, it seems safe to assume that Redrock Mimbres and Safford Hohokam were as chummy as current events allowed. Perhaps that meant hostile exclusion, but more likely the times were open and welcoming: we have the fact of Hohokam objects and ideas in Late Pithouse Mimbres to the east, and those objects and ideas had to come from somewhere. Safford was the closest source, and no mean city: the Buena Vista ball-court was one of the largest in the Hohokam region.

Buena Vista (AKA Pueblo Viejo) at Safford was at or near the extreme limit of the Hohokam ball-court system, about 240 km from Phoenix. That's about as far from Phoenix as the northernmost ball-courts at Wupatki. Presumably, life out on the edges was not tightly regulated by the center (although the size and formality of both Wupatki and Buena Vista ball-courts might give us pause). Safford, if a local political entity, must have basked in the reflected glory of Hohokam at its height: most vigorous in the Colonial period but perhaps farthest-flung in the Sedentary period. And while Mimbres people undoubtedly knew about the cities at the confluence of Salt and Gila, it's possible only a handful actually went to Phoenix, Hohokam's center. (Mary Helms' 1988 *Ulysses's Sail* suggests how these interactions may have worked: direct, politically and culturally significant, but limited in archaeological visibility.)

## Safford in the Eleventh and Twelfth Centuries

The currently emerging local sequence (Neuzil 2006 [ed.]) gives us Safford phases for the post-Hohokam, post-Mimbres era (that is, the eleventh and twelfth centuries). But, to me, this period remains one of the most mysterious and intriguing in the Upper Gila. In the Cliff and Redrock valleys of the upper Gila, it's hard to see anyone at home, in residence (Lekson 1990, 2002, 2006). Perhaps the sites are there, but survey fails to disclose them. If seems that the Uppermost Gila was empty from the late eleventh to the early thirteenth centuries. It was a time of retraction: the post-Chaco reformation in the Four Corners (Lekson 1999; Lipe 2006), and the collapse of the Sedentary Period into the Phoenix Basin Classic (Abbott 2003). This is not to say that other areas were not occupied (Tucson was; Zuni was), but the larger regional polities with centers (and therefore peripheries) were disorganizing. The abandonment of the Upper Gila (and

Mimbres if such was the case; Creel 1999 vs Shafer 1999) surely reflects those larger regional histories. What was happening in Safford? Did Safford rise as Mimbres fell, or was the valley comparably diminished? We have sites and phases, but what was happening demographically?

## Safford as Center: Salado

Salado's checkered history need not distract us here. Enough to say that an updated version of the original idea—Plateau peoples spilling into the deserts—has new life. Not so much Gladwin's conquering Pueblo hordes, but Salado as something developed from the mingling of Four Corners peoples with locals, as at Point of Pines (Haury 1958) a scant 60 km to the north of Safford; or on the San Pedro Valley (DiPeso 1958), 85 km west; or at Safford itself, most dramatically shown by Goat Hill (Woodson 1999). At Point of Pines and in the San Pedro, Plateau peoples formed enclaves among substantial local populations. That appears to be the case, too, at Safford. Who those local people might be is intriguing: there's a large dose of Plateau-based Tularosa pottery in Safford immediately preceding the flashier Salado polychromes (Lekson 1996). Did they come in waves?

In 1973, Jeff Brown suggested a central role for Safford in the formation of Salado, and certainly it is in Safford and San Simon and the Upper Gila that Salado enjoys its greatest archaeological clarity. Archaeologists no longer talk about Salado in the Phoenix Basin; the period is simply Classic Hohokam. But the term is still useful on the Upper Gila (Lekson 2002). If Safford was ever a center, it was during the late thirteenth and fourteenth centuries. What happened in Safford and the Upper Gila, and why?

## Safford as Samarqand: Crossroads of Cultures

In 1370—coincidentally, at Salado's height—Timur the Lame made Samarqand his capital. The city had a deep history as a place on the Silk Road between the Mediterranean and China. And, of course, Samarqand had local history: for example, Timur was born near there, of a locally prominent family. He was raised locally and thought globally, becoming Tamerlane, "Prince of Destruction" and ruler of a vast empire. But Samarqand was what it was—Timur's capital—because it was the most interesting of peripheries: an important place between more important places.

So too Safford, in the thirteenth through fifteenth centuries. Safford was intermediate between Classic Period Phoenix (to the northwest) and its peer polity Paquimé (to the southeast). During the thirteenth century, Paquimé rose while Phoenix fell. Phoenix's fall was not sudden or dramatic—"centuries of decline," in David Abbott's terms. Phoenix remained the oldest and in some ways the Southwest's most remarkable center through all but the end of Paquimé meteoric career. The Salado florescence in the Upper Gila may have been something like Samarqand: local, but deeply engaged with and affected by its intermediate place between two of the Southwest's most impressive ancient societies.

A straight line between Paquimé and Phoenix does not run through Safford. The direct route might (with reasonable latitude) pass through the Animas phase Paquimé outliers of New Mexico's "bootheel," and thence up the San Pedro (or perhaps through Tucson) to Casa Grande. How, then, was Safford intermediate? Consider Samarqand and the Silk Road. The Silk Road, or Route, was not a straight line or a single road. Rather, the key connection in Old World "big history" was an inter-braided mesh of a dozen parallel roads, trails, and sea routes, from the Indian Ocean on the south to Samarqand in the north. So too, the connections between Phoenix and Paquimé.

Think of Safford that way: as a node or crossroad on a web of connections between the two centers, and an important place in its own right. We know that Paquimé had interests very close to Safford: the ricolite mines of Redrock (again, a scant 80 km distant) (DiPeso 1974). Salado sites at Redrock produce impressive quantities of Paquimé pottery (Lekson 2002). It is possible that the easternmost Hohokam platform mound —signature of the Classic Period—was at the Buena Vista site (William Doelle, personal communication 2005; Tyberg 2000) (Figure Forty One).

Safford will not make sense without reference to the larger histories and geographies of its times: First, an interface between Hohokam at its height and Mimbres. Centuries later, an influx of Pueblo peoples from the north, the brilliant last gasp of Hohokam to the west, and the astonishing super-nova of Casas Grandes to the east. Safford was indeed peripheral to better-known Great Events of the ancient Southwest. But, at the interface of those events, Safford synthesized Salado—at least, the Upper Gila aspect of Salado—and, perhaps, a faint reflection of Samarqand's situation.

In 2001, Samarqand joined Chaco, Paquimé, and Mesa Verde as a World Heritage Site, officially "Samarqand—Crossroads of Cultures." Safford was never Samarqand, the capital of a mighty empire. But it might be useful to adopt Samarqand's tag-line: Safford, *Crossroads of Cultures*. In poorly known districts, the archaeologist's first charge is to build a local sequence, then to figure out the local settlement system and ecology, and finally to look for evidence of distant contacts, if any. Thus spake the Real Mesoamerican Archaeologist, thirty years ago (Flannery 1976). The temptation remains strong to understand the local locally at first, and thereafter to integrate the local with distant places and events—Safford, for example, would be isolated until proven otherwise. But that might be exactly the wrong way to understand Safford. If we assume, instead, that the Southwest was densely inter-connected (like the rest of Post-Classic North America), we will reach a rather different view of Safford. Think big: better to be Timur than timid.

# BIBLIOGRAPHY

Abbott, David R.
2000 Ceramics and Community Organization Among the Hohokam. The University of Arizona Press, Tucson.

Abbott, David R. (editor)
2003 *Centuries of Decline during the Hohokam Classic Period at Pueblo Grande*. The University of Arizona Press, Tucson.

Abbott, David R., Scott E. Ingram, and Brent G. Kober
2006 Hohokam Exchange and Early Classic Period Organization in Central Arizona: Focal Villages or Linear Communities? *Journal of Field Archaeology* 31(3):285-305.

Ackerly, Neal W.
1997 Mimbreño and Gileño Apache Irrigation Systems, 1853-1859. *Kiva* 62(4):349-363.

Adams, E. Charles
1998 Late Prehistory in the Middle Little Colorado River Area: A Regional Perspective. In *Migration and Reorganization: The Pueblo IV Period in the American Southwest*, edited by Katherine A. Spielmann, pp. 53-63. Anthropological Research Papers No. 51. Arizona State University, Tempe.

Ahlstrom, Richard V. N.
1997 Safford Valley Settlement Patterns. In *The Sanchez Copper Project: Vol. 1. Archaeological Investigations in the Safford Valley, Graham County, Arizona*, edited by Gregory R. Seymour, Richard V. N. Ahlstrom, and David P. Doak, pp. 9.1-9.10. Archaeological Report 94-82 (Revised 1997). SWCA Environmental Consultants, Tucson.

Ahlstrom, Richard V. N., Carla R. Van West, and Jeffrey S. Dean
1995 Environmental and Chronological Factors in the Mesa Verde-Northern Rio Grande Migration. *Journal of Anthropological Archaeology* 14(2):125-142.

Anthony, David W.
1990 Migration in Archaeology: The Baby and the Bathwater. *American Anthropologist* 92:895-914.

1994 Prehistoric Migration as Social Process. Paper presented at the 59th Annual Meeting of the Society for American Archaeology, Anaheim.

1997 Prehistoric Migration as Social Process. In *Migrations and Invasions in Archaeological Explanation*, edited by John Chapman and Helena Hamerow, pp. 21-32. BAR International Series 664. Hadrian Books, Oxford, U.K.

Arizona State Water Commissioner

1920 *Map of Surveys Showing Irrigated Lands Under Ditches Taking Water from Gila River or Tributaries, District No. 3*. Surveyed 1920, Scale: 1 inch = 1,000 feet. State Water Commissioner, Gila River Determination, Graham County, Arizona. On file, Office of the Gila Valley Irrigation District, Safford, Arizona.

Ayer, Mrs. Edward E. (translator)

1965 *The Memorial of Fray Alonso de Benavides*. Horn and Wallace, Albuquerque.

Babb, Cyrus C.

1901 Reconnaissance of the Gila River Basin. In "Report of Progress of Stream Measurements for the Calendar Year 1899," by F. H. Newell, pp. 334-358. Annual Reports of the United States Geological Survey 21(4). Government Printing Office, Washington, D.C.

Ball, Eve

1970 *In the Days of Victorio: Recollections of a Warm Springs Apache*. The University of Arizona Press, Tucson.

Barnard, Alan

1983 Contemporary Hunter-Gatherers: Current Theoretical Issues in Ecology and Social Organization. *Annual Review of Anthropology*, Vol. 12, 1983 (1983):193-214.

Barth, Fredrik

1969a Introduction: Ethnic Groups and Boundaries. In *Ethnic Groups and Boundaries: The Social Organization of Culture Difference*, edited by Fredrik Barth, pp. 9-38. Little, Brown, and Co., Boston.

1969b Ecological Relationships of Ethnic Groups in Swat, North Pakistan. In *Environment and Cultural Behavior: Ecological studies in Cultural Anthropology*, edited by Andrew P. Vayda, pp. 362-376. American Museum Sourcebooks in Anthropology, The Museum of Natural History. The Natural History Press, Garden City, New York.

1994 Enduring and Emerging Issues in the Analysis of Ethnicity. In *The Anthropology of Ethnicity: Beyond 'Ethnic Groups and Boundaries'*, edited by Hans Vermeulen and Cora Govers, pp. 11-32. Het Spinhuis, Amsterdam.

Bandelier, Adolph F.

1892 *Final Report of Investigations Among the Indians of the Southwestern United States, Carried on Mainly in the Years from 1880 to 1885, Part II.* Papers of the Archaeological Institute of America, American Series IV. University Press, Cambridge.

Basso, Keith H.

1996 *Wisdom Sits in Places: Landscape and Language Among the Western Apache.* University of New Mexico Press, Albuquerque.

Baugh, Timothy G., and Mark Sechrist

2001 *Protohistoric Apachean Adaptations within the Basin and Range Province of South-Central New Mexico and West Texas: A Perspective from the Fort Bliss Reservation.* TRC, El Paso.

Benedict, Ruth

1934 *Patterns of Culture.* Houghton Mifflin, Boston.

Bentley, G. Carter

1987 Ethnicity and Practice. *Comparative Studies in Society and History* 29(1):24-55.

Bernardini, Wesley

1998 Conflict, Migration, and the Social Environment: Interpreting Architectural Change in Early and Late Pueblo IV Aggregations. In *Migration and Reorganization: The Pueblo IV Period in the American Southwest*, edited by Katherine A. Spielmann, pp. 91-114. Anthropological Research Papers No. 51. Arizona State University, Tempe.

2002 The Gathering of the Clans: Understanding Ancestral Hopi Migration and Identity, A.D. 1275-1400. Unpublished Ph.D. Dissertation, Department of Anthropology, Arizona State University, Tempe.

2005 Reconsidering Spatial and Temporal Aspects of Prehistoric Cultural Identity: A Case Study from the American Southwest. *American Antiquity* 70(1):31-54.

Berry, Michael S.

1982 *Time, Space, and Transition in Anasazi Prehistory.* University of Utah Press, Salt Lake City.

Bettinger, Robert L., and Jelmer Earkens
1999 Point Typologies, Cultural Transmission, and the Spread of Bow-and-Arrow Technology in the Prehistoric Great Basin. *American Antiquity* 64(2):243-264.

Betzinez, Jason and Wilbur Sturtevant Nye
1959 *I Fought With Geronimo.* University of Nebraska Press, Lincoln and London.

Birkeland, Nina M.
2000 Forced migration and *deslocados* in the Huambo Province, Angola. *Norwegian Journal of Geography* 54:110-115.

Bolton, Herbert Eugene
1916 *Spanish Exploration in the Southwest, 1542-1706.* Charles Scribner's Sons, Inc., New York.
1991[1949] *Coronado: Knight of Pueblos and Plains.* University of New Mexico Press, Albuquerque.

Borden, Charles
1952 Results of Archaeological Investigations in Central British Columbia. *Anthropology in British Columbia* 3:31-43.

Bostwick, Todd W.
2002 *Landscape of the Spirits.* The University of Arizona Press, Tucson.

Botsford, Manton, and Gay Kinkade
1993 Cultural Resources Assessment of the BLM Safford District Office Project. MS. on file, Bureau of Land Management, Safford District, Arizona.

Bowden, M. J.
1992 The invention of American tradition. *Journal of Historical Geography* 18:326.

Brasher, Nugent
2007 The Chichilticale Camp of Francisco Vasquez de Coronado: The Search for the Red House. *New Mexico Historical Review* 82(4):433-468.

Brettell, Caroline B., and James F. Hollifield
2000 Migration Theory: Talking Across Disciplines. In *Migration Theory: Talking Across Disciplines*, edited by Caroline B. Brettell and James F. Hollifield, pp. 1-26. Routledge, New York.

Brown, Jeffrey L.
1973 The Origin and Nature of Salado: Evidence from the Safford Valley, Arizona. Unpublished Ph.D. dissertation, Department of Anthropology, University of Arizona, Tucson.

Bureau of Land Management
1973 *Safford, Arizona-New Mexico*: 30 by 60 minute series (1:100,000-scale) metric planimetric map. Bureau of Land Management, U.S. Department of the Interior, Washington, D.C.

Burmeister, Stefan
2000 Archaeology and Migration: Approaches to an Archaeological Proof of Migration. *Current Anthropology* 41(4):539-567.

Burrus, Ernest J. (editor)
1969 *A History of the Southwest* by Adolph F. Bandelier, Volume I and supplement. Sources and Studies for the History of the Americas: Volume VII. Jesuit Historical Institute, St. Louis University, St. Louis.

Burton, Jeffery F.
1988 Hunters and the Hunted: Prehistoric Art of Tom Ketchum Cave. *The Kiva* 53:335–356.

Bye, Robert A., Jr., and Rita Shuster
1984 Developing an Integrated Model for Contemporary and Archaeological Agricultural Subsistence Systems. In *Prehistoric Agricultural Strategies in the Southwest*, edited by S. K. Fish and P. R. Fish, pp. 147-159. Anthropological Research Papers 33. Arizona State University, Tempe.

Cameron, Catherine M.
1995 Migration and the Movement of Southwestern Peoples. *Journal of Anthropological Archaeology* 14(2):104-124.
2005 Exploring Archaeological Cultures in the Northern Southwest: What Were Chaco and Mesa Verde? *Kiva* 70(3):227-254.

Caran, S. Christopher, and James A. Neely
2006 Hydraulic Engineering in Prehistoric Mexico. *Scientific American* 295(4):78-85.

Ciolek-Torrello, Richard
1995 Early Formative Period Chronology for the Tucson Basin. *Kiva* 60(4):481-530.

Clark, Jeffery J.
2001 *Tracking Prehistoric Migration: Pueblo Settlers Among the Tonto Basin Hohokam*. Anthropological Papers of the University of Arizona, Number 65. The University of Arizona Press, Tucson.
2006 Irrigation and Migration: Bits and Pieces of Safford Basin Archaeology. *Archaeology Southwest* 20(2):6.

Clark, Jeffrey J. [editor]
2004 *Ancient Farmers of the Safford Basin: Archaeology of the U.S. 70 Safford-to-Thatcher Project.* Anthropological Papers No. 39. Center for Desert Archaeology, Tucson.

Clark, Jeffrey L., Anna A, Neuzil, and Patrick D. Lyons
2004 Safford Basin Past and Present. In *The Ancient Farmers of the Safford Basin: Archaeology of the U.S. 70 Safford-to-Thatcher Project,* edited by Jeffrey L. Clark, pp. 207-233. Anthropological Papers No. 39. Center for Desert Archaeology, Tucson.

Clark, Jeffrey. J., Fred L. Nials, and James. M. Vint
2004 Introduction. In *The Ancient Farmers of the Safford Basin: Archaeology of the U.S. 70 Safford-to-Thatcher Project,* edited by J. L. Clark, pp. 1-25. Anthropological Papers No. 39. Center for Desert Archaeology, Tucson.

Clark, T. C.
1998 Faunal Resource Depletion and Nutritional Stress in the Pueblo IV Period. In *Migration and Reorganization: The Pueblo IV Period in the American Southwest*, edited by Katherine A. Spielmann, pp. 193-207. Arizona State University, Tempe.

Cleland, Charles E.
1976 The Focal-Diffuse Model: An Evolutionary Perspective on the Prehistoric Cultural Adaptations of the Eastern United States. *Midcontinental Journal of Archaeology* 1:59-76.

Clements, F. E.
1904 Studies in the Vegetation of the State, III. The Development and Structure of Vegetation. *The Seminar*. University of Nebraska, Lincoln.

Coe, Michael D., and Kent V. Flannery
1964 Microenvironments and Mesoamerican prehistory. *Science* 143(3607):650-654.

Coffey, Grant
2006 Reevaluating Regional Migration in the Northern San Juan During the Late Pueblo I Period: A Reconnaissance Survey of the East Dove Creek Area. *Kiva* 72(1)55-70.

Cole, D.C.
1988 *The Chiricahua Apache: From War to Reservation 1846 - 1876.* University of New Mexico Press, Albuquerque.

Cole, Sally J.
1990 *Legacy on Stone: Rock Art of the Colorado Plateau and Four Corners Region.* Johnson Books, Boulder.

Colton, Harold S.

1939 *Prehistoric Culture Units and Their Relationships in Northern Arizona*. Bulletin No. 17. Museum of Northern Arizona, Flagstaff.

Colvin, Verna Rae

1997 Building Canals on the Gila River. In *Link the Past with the Present: 1997 Symposium Papers*, pp. 4-18, Graham County Historical Society, Safford.

1998 First Came the Water, and then the People: History of Water in Graham County. Unpublished photocopied manuscript, Bureau of Land Management, Safford District, Safford.

Colvin, Verna Rae, and Patricia A. Cook

2006 Historic Canals of the Safford Basin. *Archaeology Southwest* 20(2):9-11.

Comas d'Argemir, Dolors, and Joan J. Pujadas

1999 Living In/On the Frontier: Migration, Identities, and Citizenship in Andorra. *Social Anthropology* 7(3):253-264.

Cordell, Linda S.

1984 *Prehistory of the Southwest*. Academic Press, Orlando and San Diego.

Craib, Ian

1998 *Experiencing Identity*. Sage Publications, London.

Crary, Joseph S.

1997 The Chronology and Cultures of Upper (Northern) Southeast Arizona: The Formative and Classic Periods. MS prepared for the symposium The Archaeology of a Land Between: Regional Dynamics in the Prehistory and History of Southeast Arizona, sponsored by the Amerind Foundation, October 12-17, 1997, Dragoon.

Creel, Darrell

1999 The Black Mountain Phase in the Mimbres Area. In *The Casas Grandes World*, edited by Curtis R. Schaafsma and Carroll L. Riley, pp 107-120. University of Utah Press, Salt Lake City.

Creel, Darrell, and Roger Anyon

2003 New Interpretations of Mimbres Public Architecture and Space: Implications for Cultural Change. *American Antiquity* 68(1):67-92.

Cribb, Roger

1991 *Nomads in Archaeology*. Cambridge University Press, Cambridge.

Crown, Patricia L.
1984 Prehistoric Agricultural Technology in the Salt-Gila Basin. In *Hohokam Archaeology Along the Salt-Gila Aqueduct, Central Arizona Project, Vol. VII: Environment and Subsistence*, edited by Lynn S. Teague and Patricia L. Crown, pp. 207-260. Archaeological Series 150. Cultural Resources Management Section, Arizona State Museum, The University of Arizona, Tucson.
1987 Classic Period Hohokam Settlement and Land Use in the Casa Grande Ruins Area, Arizona. *Journal of Field Archaeology* 14:147-162.
1991 Craft Arts of the Hohokam. In *The Hohokam: Ancient People of the Desert*, edited by David G. Noble, pp. 39-61. School of American Research Press, Santa Fe.
1994 *Ceramics and Ideology: Salado Polychrome Pottery*. University of New Mexico Press, Albuquerque.

Cuba, Lee, and David M. Hummon
1993 Constructing a Sense of Home: Place Affiliation and Migration Across the Life Cycle. *Sociological Forum* 8(4):547-572.

Damp, Jonathan E., Stephen A. Hall, and Susan J. Smith
2002 Early Irrigation of the Colorado Plateau near Zuni Pueblo, New Mexico. *American Antiquity* 67(4):665-676.

Danson, Edward B.
1957 *An Archaeological Survey of West Central New Mexico and East Central Arizona*. Papers of the Peabody Museum of Archaeology and Ethnology, Vol. XLIV, No. 1. Cambridge.

Dean, Jeffrey S.
1996 Kayenta Anasazi Settlement Transformations in Northeastern Arizona: A.D. 1150-1350. In *The Prehistoric Pueblo World: A.D. 1150-1350*, edited by Michael A. Adler, pp. 29-47. The University of Arizona Press, Tucson.
2000 Introduction: The Salado Concept. In *Salado*, edited by Jeffrey S. Dean, pp. 3-16. Amerind Foundation New World Studies Series Number Four. Amerind Foundation and University of New Mexico Press, Albuquerque.

Dean, Jeffrey S., William H. Doelle, and Janet D. Orcutt
1994 Adaptive Stress: Environment and Demography. In *Themes in Southwest Prehistory*, edited by George J. Gumerman, pp. 53-86. School of American Research Press, Santa Fe.

Díaz-Andreu, Margarita
1996 Constructing Identities Through Culture: The Past in the Forging of Europe. In *Cultural Identity and Archaeology: The Construction of European Communities*, edited by Paul Graves-Brown, Sian Jones, and Clive Gamble, pp. 48-61. Routledge, London.

Diehl, Michael W.
2004 Plant Remains and Resource Use in the Safford Basin. In *Ancient Farmers of the Safford Basin: Archaeology of the U.S. 70 Safford-to-Thatcher Project*, edited by Jeffery J. Clark, pp. 147-164. Anthropological Papers 39. Center for Desert Archaeology, Tucson.

Di Peso, Charles C.
1956 *The Upper Pima of San Cayetana del Tumacacari: An Archaeohistorical Reconstruction of the Ootam of Pimeria Alta.* Paper No. 7. Amerind Foundation, Dragoon, Arizona.
1958 *The Reeve Ruin of Southeastern Arizona: A Study of a Prehistoric Western Pueblo Migration into the Middle San Pedro Valley.* Paper No. 8. Amerind Foundation, Dragoon, Arizona.
1974 *Casas Grandes: A Fallen Trading Center of the Gran Chichimeca.* Vols 1-3. Publication No. 9. Amerind Foundation, Dragoon, Dragoon
1979 Prehistory: O'otam. In *Southwest*, edited by Alfonso Ortiz, pp. 91-99. Handbook of North American Indians 9. Smithsonian Institution, Washington, D.C.

Dobyns, Henry F.
1966 An Appraisal of Techniques with a New Hemispheric Estimate. *Current Anthropology* 7(4):395-416.

Doolittle, Christopher J.
2000 Petroglyph Research at Antelope Hill. In *Of Stones and Spirits: Pursuing the Past at Antelope Hill*, edited by Joan S. Schneider and Jeffrey H. Altschul, pp. 83–118. Technical Series 76. Statistical Research, Inc., Tucson.

Doolittle, William E.
1984 Agricultural Change as an Incremental Process. *Annals of the Association of American Geographers* 74:124-138.
1988 *Pre-Hispanic Occupance in the Valley of Sonora, Mexico: Archaeological Confirmation of Early Spanish Reports.* Anthropological Papers of the University of Arizona, Number 48. The University of Arizona Press, Tucson.

1990 *Canal Irrigation in Prehistoric Mexico: The Sequence of Technological Change*. University of Texas Press, Austin.

2000 *Cultivated Landscapes of Native North America*. Oxford Geographical and Environmental Series. Oxford University Press, Oxford, U.K..

2006 Agricultural Manipulation of Floodplains in the Southern Basin and Range Province. *Catena* 65:179-199.

2008 Landscapes, Locales, Fields, and Food. In *The Archaeology of a Land Between: Regional Dynamics in the Prehistory and History of Southeastern Arizona*. Archaeological Series No. 12. Amerind Foundation, Dragoon, Arizona. Manuscript submitted for publication.

Doolittle, William E., and Jonathan B. Mabry

2006 Environmental Mosaics, Agricultural Diversity, and the Evolutionary Adoption of Maize in the American Southwest. In *Histories of Maize: Multidisciplinary Approaches to the Prehistory, Linguistics, Biogeography, Domestication, and Evolution of Maize*, edited by J. E. Stalter, R. H. Tykot, and B. F. Benz, pp. 109-121. Academic Press and Elsevier, Amsterdam.

Doolittle, William E., and James A. Neely (editors)

2004 *The Safford Valley Grids: Prehistoric Cultivation in the Southern Arizona Desert*. Anthropological Papers of the University of Arizona, Number 70. The University of Arizona Press, Tucson.

Doyel, David E.

2000 In Pursuit of Salado in the Sonoran Desert. In *Salado*, edited by Jeffrey S. Dean, pp. 295–314. Amerind Foundation New World Studies Series Number Four. Amerind Foundation Publications, Dragoon, Arizona, and University of New Mexico Press, Albuquerque.

Eastern Arizona College (EAC)

1981 *The Daley Site: An Archaeological Salvage Report*. Museum of Anthropology Publication Number 1. Eastern Arizona College, Thatcher, Arizona.

Elliot, Henry

2002 *Imperial Spain: 1469-1716*. Penguin Books, New York and London.

Ezzo, Joseph A.
2005 Site and Community Structure at Sleeping Snake Village and Los Venados. In *Ballcourt on the Bajada: Data Recovery at Sleeping Snake Village (AZ BB:9:104[ASM]) and Los Venados (AZ BB:9:186[ASM]), Oro Valley, Arizona*, edited by Joseph A. Ezzo, pp. 795–810. Cultural Resources Report No. 05–290, SWCA Environmental Consultants, Tucson.

Ezzo, Joseph A. (editor)
2007 *Sojourns in the Foothills: The Dos Pobres/San Juan Data Recovery Project, Volume 1: Native American Sites*, edited by Joseph A. Ezzo, pp. 197–242. Cultural Resources Report No. 06–191. SWCA Environmental Consultants, Tucson.

Ezzo, Joseph A., and Jeffrey H. Altschul
1993a An Archaeological Survey of Palo Verde Point, Imperial County, California: A Class III Cultural Resources Survey and Evaluation. In *Glyphs and Quarries of the Lower Colorado River Valley,* compiled by Joseph A Ezzo and Jeffrey H. Altschul, Part 5. Technical Series No. 44. Statistical Research, Inc., Tucson.

1993b An Archaeological Survey of Pilot Knob, Imperial County, California: A Class III Cultural Resources Survey and Evaluation. In *Glyphs and Quarries of the Lower Colorado River Valley,* compiled by Joseph A Ezzo and Jeffrey H. Altschul, Part 3. Technical Series No. 44. Statistical Research, Inc., Tucson.

Ferguson, T. J.
1996 *Historic Zuni Architecture and Society: An Archaeological Application of Space Syntax.* Anthropological Papers of the University of Arizona, Number 60. The University of Arizona Press, Tucson.

Fewkes, Jesse W.
1898 A Preliminary Account of Archaeological Field Work in Arizona 1897. In *Annual Reports of the Smithsonian Institution for 1897*, pp. 601-623. Government Printing Office, Washington, D.C.

1904 Two Summer's Work in Pueblo Ruins. In *Twenty-Second Annual Report of the Bureau of American Ethnology, 1900-1901, Part 1*, pp. 3-196. Smithsonian Institution, Washington, D.C.

Flannery, Kent V. (editor)
1976 *The Early Mesoamerican Village.* Academic Press, New York.

Flint, Richard and Shirley Cushing Flint
2005 *Documents of the Coronado Expedition, 1539-1541: "They Were Not Familiar with His Majesty nor Did They Wish to Be His Subjects,"* Southern Methodist University Press, Dallas.
Forbes, Jack D.
1960 *Apache, Navajo, and Spaniard.* University of Oklahoma, Norman.
Ford, James A.
1969 *A Comparison of Formative Cultures in the Americas: Diffusion of the Psychic Unity of Man.* Smithsonian Institution Press, Washington, D.C.
Frayne, Bruce, and Wade Pendelton
2001 Home is Where the Heart Is: Namibians on Cross-Border Migration and Regional Integration. *African Studies* 60(2):205-224.
Galloway, Patricia
1991 The Archaeology of Ethnohistorical Narrative. In *Columbian Consequences, Volume 3, The Spanish Borderlands in Pan-American Perspective*, edited by David H. Thomas, pp. 453-469. Smithsonian Institution Press, Washington and London.
Gardner, Andrew
2002 Social identity and the duality of structure in late Roman-period Britain. *Journal of Social Archaeology* 2(3):323-351.
Gelles, Paul H.
1996 The Political Ecology of Irrigation in an Andean Peasant Community. In *Canals and Communities: Small-Scale Irrigation Systems*, edited by Jonathan B. Mabry, pp. 88-115. The University of Arizona Press, Tucson.
Gilman, Patricia A.
1997 *Wandering Villagers: Pit Structures, Mobility and Agriculture in Southeastern Arizona.* Anthropological Research Papers No. 49. Arizona State University, Tempe.
Gladwin, Harold Sterling
1957 *A History of the Ancient Southwest.* Bond Wheelwright, Portland, Maine.
Gladwin, Winifred, and Harold S. Gladwin
1929 *The Red-on-buff Culture of the Gila Basin.* Medallion Papers No. 3. Gila Pueblo, Globe, Arizona.

Goodwin, Grenville and Keith H. Basso
1971 *Western Apache Raiding and Warfare.* The University of Arizona Press, Tucson.

Gregory, David A.
1981 Western Apache Archaeology: Problems and Approaches. In *The Protohistoric Period in the North American Southwest. AD 1450–1700*, edited by David R. Wilcox and William B. Masse, pp. 257-274. Anthropological Research Papers No. 24. Arizona State University, Tempe.
1995 Research Orientation and Research Problems: Prehistoric Sites. In *The San Carlos Reservoir Cultural Resources Survey, Volume 1*, edited by Andrew T. Black and Margerie Green, pp. 33-42. Cultural Resources Report No. 87. Archaeological Consulting Services, Ltd., Tempe.
2001 An Evaluation of Early Agricultural Period Chronology in the Tucson Basin. In *Excavations in the Santa Cruz River Floodplain: The Early Agricultural Period Component at Los Pozos*, edited by David A. Gregory, 237-254. Anthropological Papers No. 21. Center for Desert Archaeology, Tucson.

Gunnerson, Dolores A.
1956 The Southern Athabascans: Their Arrival in the Southwest. *El Palacio*, December: 347-365.

Gunnerson, James H.
1960 *An Introduction to Plains Apache Archeology—The Dismal River Aspect.* Bureau of American Ethnology, Bulletin 173. Anthropological Papers No. 58. Smithsonian Institution, Washington, D.C.

Gunnerson, James H., and Dolores A. Gunnerson
1971 *Apachean Culture, History and Ethnology.* Anthropological Papers of the University of Arizona, Number 21. The University of Arizona Press, Tucson.

Haggerstrand, T.
1967 *Innovation Diffusion as a Spatial Process.* University of Chicago Press, Chicago.

Hall, Susan D., and Jeffery J. Clark
2004 AZ CC:2:289(ASM). In *Ancient Farmers of the Safford Basin: Archaeology of the U.S. 70 Safford-to-Thatcher Project*, edited by Jeffery J. Clark, pp. 23-41. Anthropological Papers No. 39. Center for Desert Archaeology, Tucson.

Hammond, George Peter and Agapito Rey
1929 *Expedition into New Mexico Made by Antonio De Espejo 1582-1583*. The Quivira Society, Los Angeles.
1940 *Narratives of the Coronado Expedition 1540-1542*. University of New Mexico Press, Albuquerque.
1966 *The Rediscovery of New Mexico, 1580-1594, The Explorations of Chamuscado, Espejo, Castano de Sosa, Morlete, and Leyva de Bonilla and Humana*. University of New Mexico Press, Albuquerque.

Harlan, Mark. E.
2007 Historians and Archaeologists: Proposals for Connecting in a Common Past. *New Mexico Historical Review*. 82(4):501-540.

Harrington, John P.
1940 *Southern Peripheral Athapaskan Origins, Divisions, and Migrations*. Smithsonian Miscellaneous Collections, Vol. 100, pp. 503-532. Washington, D.C.

Hartmann, William K., and Richard Flint
2001 Migrations in Late Anasazi Prehistory: "Eyewitness" Testimony. *Kiva* 66(3):375-386.

Haug, Ruth
2002 Forced Migration, Processes of Return and Livelihood Construction among Pastoralists in Northern Sudan. *Disasters* 26(1):70-84.

Haury, Emil W.
1936 *The Mogollon Culture of Southwestern New Mexico*. Medallion Papers No. 20. Gila Pueblo, Globe, Arizona.
1958 Evidence at Point of Pines for a Prehistoric Migration from Northern Arizona. In *Migrations in New World Culture History*, edited by Raymond H. Thompson, pp. 1-6. University of Arizona Social Science Bulletin No. 27. The University of Arizona Press, Tucson.
1976 *The Hohokam: Desert Farmers and Craftsmen. Excavations at Snaketown, 1964-1965*. The University of Arizona Press, Tucson.

Hayden, Julian D.
1972 Hohokam Petroglyphs of the Sierra Pinacate, Sonora and the Hohokam Shell Expeditions. *The Kiva* 37:74–83.

Hays–Gilpin, Kelley
2004 *Ambiguous Images*. AltaMira Press, Walnut Creek, California.

Hegmon, Michelle
2003 Setting Theoretical Egos Aside: Issues and Theory in North American Archaeology. *American Antiquity* 68(2):213-244.

Heidke, James M.
1999 Cienega Phase Incipient Plainware from Southeastern Arizona. *Kiva* 64(3):311-338.

Heleniak, Timothy
2004 Migration of the Russian Diaspora After the Breakup of the Soviet Union. *Journal of International Affairs* 57(2):99-117.

Helms, Mary W.
1988 *Ulysses' Sail: An Ethnographic Odyssey of Power, Knowledge, and Geographical Distance.* Princeton University Press, Princeton.

Herr, Sarah A.
2001 *Beyond Chaco: Great Kiva Communities on the Mogollon Rim Frontier.* Anthropological Papers of the University of Arizona, Number 66. The University of Arizona Press, Tucson.

Herr, Sarah, and Jeffrey J. Clark
1997 Patterns in the Pathways: Early Historic Migrations in the Rio Grande Pueblos. *Kiva* 62(4):365-389.

Hesse, India S.
2007 Flaked and Ground Stone Artifacts. In *Sojourns in the Foothills: The Dos Pobres/San Juan Data Recovery Project, Volume 1: Native American Sites*, edited by Joseph A. Ezzo, pp. 197–242. Cultural Resources Report No. 06–191. SWCA Environmental Consultants, Tucson.

Hillier, Bill, and Julienne Hanson
1984 *The Social Logic of Space.* Cambridge University Press, Cambridge.

Hodge, Frederick Webb (editor)
1990 The Narrative of Alvar Nunez Cabeza de Vaca. In *Spanish Explorers in the Southern United States 1528-1543*, pp. 1-123. The Texas State Historical Association, Austin.

Hough, Walter
1907 *Antiquities of the Upper Gila and Salt River Valleys in Arizona and New Mexico.* Bureau of American Ethnology Bulletin 35. Smithsonian Institution, Washington.

Howard, Jerry B.
2006 Hohokam Irrigation Communities: A Study of Internal Structure, External Relationships and Sociopolitical

Complexity. Unpublished Ph.D. dissertation, School of Human Evolution and Social Change, Arizona State University, Tempe.

Howard, Jerry B., and Gary Huckleberry

1991 *The Operation and Evolution of an Irrigation System: The East Papago Canal Study*. Publications in Archaeology 18. Soil Systems, Inc., Phoenix.

Huckell, Bruce B.

1995 *Of Mashes and Maize: Preceramic Agricultural Settlements in the Cienega Valley, Southeastern Arizona*. Anthropological Papers, Number 59. The University of Arizona Press, Tucson.

Huckleberry, Gary

2005 Geomorphology of the Safford Basin and Canal Studies at Epley's Ruin, AZ CC:2:64 (ASM). In Archaeological Investigations along US 70 and State Route 75 from Solomon to Apache Grove, Graham and Greenlee Counties, Southeast Arizona, Vol. 2, Epley's Ruin (AZ CC:2:64[ASM]): A Long-Lived Irrigation Community in the Safford Basin from 200 B.C. to A.D. 1450, edited by Annick Lascaux and Barbara K. Montgomery. Tierra Archaeological Report No. 2005-94. Tierra Right of Way Services, Ltd., Tucson. Draft report in review.

Hunt, Robert C.

1994 Irrigation in Cuicatlán: The Question of Río Grande Waters. In *Caciques and Their People: A Volume in Honor of Ronald Spores*, edited by J. Marcus and J. F. Zeitlin, pp. 163-187. Museum of Anthropology Anthropological Papers 89. University of Michigan, Ann Arbor.

Hunt, Robert C., David Guillet, David R. Abbott, James Bayman, Paul Fish, Suzanne Fish, Keith Kintigh, and James A. Neely

2005 Plausible Ethnographic Analogies for the Social Organization of Hohokam Canal Irrigation. *American Antiquity* 70(3):433-456.

Ingold, T.

1987 *Hunters, Pastoralists and Ranchers*. Cambridge Studies in Social Anthropology. Cambridge University Press, London.

Irwin-Williams, Cynthia

1985 Review of *Time, Space, and Transition in Anasazi Prehistory*, by Michael S. Berry (1982). *The Kiva* 51(1):44-47.

Jenkins, Richard

1996 *Social Identity*. Routledge, London.

Jernigan, Wes
1993 Foreword. In *The Owens-Colvin Site of the Safford Valley*, by Pam Rule, pp. 5. Museum of Anthropology Publication Number 3. Eastern Arizona College, Thatcher.

Jett, Stephen C.
1964 Pueblo Indian Migrations: An Evaluation of the Possible Physical and Cultural Determinants. *American Antiquity* 29:281-300.

John, Elizabeth A. H.
1989 *Views From the Apache Frontier, Report on the Northern Provinces of New Spain.* By José Cortés, edited by Elizabeth A.H. John; Translated by John Wheat. University of Oklahoma Press, Norman.

Johnson, Alfred E., and William W. Wasley
1966 Archaeological Excavations near Bylas, Arizona. *The Kiva* 31(4):205-253.

Jones, Gill
1999 'The Same People in the Same Places'? Socio-Spatial Identities and Migration in Youth. *Sociology* 33(1):1-22.

Jones, Terry L., and Kathryn A. Klar
2005 Diffusionism Reconsidered: Linguistic and Archaeological Evidence for Prehistoric Polynesian Contact with Southern California. *American Antiquity* 70(3):457-484.

Kamen, Henry
2003 *Empire: How Spain Became a World Power, 1492-1763.* HarperCollins Publishers, New York.

Khazanov, A.M.
1984 *Nomads and the Outside World.* Cambridge University Press, Cambridge.

Kidder, Alfred Vincent
1975 *An Introduction to the Study of Southwestern Archaeology,* Eighth Edition. Yale University Press, New Haven.

Kohler, Timothy A., George J. Gumerman, and Robert G. Reynolds
2005 Simulating Ancient Societies: Computer Model is Helping Unravel the Archaeological Mysteries of the American Southwest. *Scientific American* 293(1):76-84.

Lascaux, Annick and Gary Huckleberry
2006 Epley's Ruin. *Archaeology Southwest* 20(2):15.

Lascaux, Annick and Barbara K. Montgomery (editors)
2005 Archaeological Investigations along US 70 and State Route 75 from Solomon to Apache Grove, Graham and Greenlee Counties, Southeast Arizona, Vol. 2, Epley's Ruin (AZ CC:2:64[ASM]): A Long-Lived Irrigation Community in the Safford Basin from 200 B.C. to A.D. 1450. Tierra Archaeological Report No. 2005-94. Tierra Right of Way Services, Ltd., Tucson. Draft report in review.

Laurance, W. F., R. K. Didham, and M. E. Powers
2001 Ecological Boundaries: A Search for Synthesis. *Trends in Ecological Evolution* 16:70-71.

Lekson, Stephen H.
1986 Mimbres Riverine Adaptations. In *Mogollon Variability*, edited by C. Benson and S. Upham, pp 181-190. Occasional Paper 15. New Mexico State University Museum, Las Cruces.
1990 *Mimbres Archaeology of the Upper Gila, New Mexico*. Anthropological Papers of the University of Arizona, Number 30. The University of Arizona Press, Tucson.
1993 Chaco, Hohokam and Mimbres: The Southwest in the 11th and 12th Centuries. *Expedition* 35(1):44-52.
1996 Southwestern New Mexico and Southeastern Arizona, A.D. 900 to 1300. In *The Prehistoric Pueblo World, A.D. 1150-1350*, edited by Michael A. Adler, pp. 170-176. The University of Arizona Press, Tucson.
1999 *The Chaco Meridian: Centers of Political Power in the Ancient Southwest*. Altamira Press, Walnut Creek, California.
2000 Salado in Chihuahua. In *Salado*, edited by Jeffrey S. Dean, pp. 275-294. Amerind Foundation New World Series Number Four. American Foundation Publications, Dragoon, Arizona, and University of New Mexico Press, Albuquerque.
2002 *Salado Archaeology of the Upper Gila, New Mexico*. Anthropological Papers of the University of Arizona, Number 67. The University of Arizona Press, Tucson.
2006 *The Archaeology of the Mimbres Region, Southwestern New Mexico, U.S.A.* British Archaeological Reports International Series 1466. Archaeopress, Oxford, U.K..
2008 A History of the Ancient Southwest. SAR Press, Santa Fe. In press.

Lekson, Stephen H., Michael Bletzer, and A. C. MacWilliams
2004 Pueblo IV in the Chihuahuan Desert. In *The Protohistoric Pueblo World, A.D. 1275-1600*, edited by E. Charles Adams and Andrew I. Duff, pp 53-61. The University of Arizona Press, Tucson.

Lekson, Stephen H., and Catherine M. Cameron
1995 The Abandonment of Chaco Canyon, the Mesa Verde Migrations, and the Reorganization of the Pueblo World. *Journal of Anthropological Archaeology* 14(2):184-202.

Lekson, Stephen H., Curtis P. Nepstad-Thornberry, Brian E. Yunker, Toni S. Laumbach, David P. Cain, and Karl W. Laumbach
2002 Migrations in the Southwest: Pinnacle Ruin, Southwestern New Mexico. *Kiva* 68(2):73-101.

Levy, Thomas E., and Augustin F. C. Holl
2002 Migrations, Ethnogenesis, and Settlement Dynamics: Israelites in Iron Age Canaan and Shuwa-Arabs in the Chad Basin. *Journal of Anthropological Archaeology* 21:83-118.

Lightfoot, Kent G.
1978 Population Movements and Social Interaction in the Prehistoric Southwest: An Example from Springerville, Arizona. In *Analytical Approach to Cultural Resource Management: The Little Colorado Planning Unit*, edited by Fred Plog, pp. 187-199. Arizona State University Anthropological Research Papers No. 13 and USDA Forest Service Cultural Resources Report No. 19. Arizona State University, Tempe, and USDA Forest Service, Springerville.
1995 Culture Contact Studies: Redefining the Relationship between Protohistoric and Historical Archaeology. *American Antiquity* 60(2):199-217.

Lilley, Ian
2004 Diaspora and Identity in Archaeology: Moving Beyond the Black Atlantic. In *A Companion to Social Archaeology*, edited by Lynn Meskell and Robert W. Preucel, pp. 287-312. Blackwell Publishing, Malden.

Lindsay, Alexander Johnston, Jr.
1987 Anasazi Population Movements to Southeastern Arizona. *American Archaeology* 6(3):190-198.

Lipe, William D.

1995 The Depopulation of the Northern San Juan: Conditions in the Turbulent 1200s. *Journal of Anthropological Archaeology* 14(2):143-169.

2006 Notes from the North. In *Archaeology of Chaco Canyon*, edited by Steven H. Lekson. SAR Press, Santa Fe.

Lucassen, Jan, and Leo Lucassen

1997 Migration, Migration History, History: Old Paradigms and New Perspectives. In *Migration, Migration History, History: Old Paradigms and New Perspectives*, edited by Jan Lucassen and Leo Lucassen, pp. 9-38. Peter Lang, Bern.

Lyman, R. Lee and Michael J. O'Brien

2001 The Direct Historical Approach, Analogical Reasoning, and Theory in Americanist Archaeology. *Journal of Archaeological Method and Theory* 8(4):303-342.

Lyons, Patrick D.

2003 *Ancestral Hopi Migrations*. Anthropological Papers of the University of Arizona, No. 68. The University of Arizona Press, Tucson.

Lyons, Patrick D., and Alexander J. Lindsay, Jr.

2006 Perforated Plates and the Salado Phenomenon. *Kiva* 72(1):5-54.

Lyons, Patrick D., Elizabeth J. Miksa, Sergio F. Castro-Reino, Carlos Lavayen

2005 Using Petrography to Demonstrate the Link Between Roosevelt Red Ware and Ancient Immigrants. Paper presented at the 70th Annual Meeting of the Society for American Archaeology, Salt Lake City, Utah.

Mabry, Jonathan B.

1998 Conclusion. In *Archaeological Investigations of Early Village Sites in the Middle Santa Cruz Valley: Analyses and Synthesis, Part 2*, edited by Jonathan B. Mabry, pp. 757-792. Anthropological Papers No. 19. Center for Desert Archaeology, Tucson.

1999 Changing Concepts of the First Period of Agriculture in the Southern Southwest. *Archaeology Southwest* 13(1):3.

Mabry, Jonathan B. (editor)

2007 Las Capas: Early Irrigation and Sedentism in a Southwestern Floodplain. Anthropological Papers 28. Center for Desert Archaeology, Tucson. Draft.

MacWilliams, A. C., and Jane H. Kelley

2004 A Ceramic Boundary in Central Chihuahua. In *Surveying the Archaeology of Northwest Mexico*, edited by, Gillian E. Newell and Emiliano Gallaga, pp. 247-264. University of Utah Press, Salt Lake City.

Malville, Nancy J.

2001 Long-Distance Transport of Bulk Goods in the Pre-Hispanic American Southwest. *Journal of Anthropological Anthropology* 20:230-243.

Marshall, Joan, and Natalie Foster

2002 "Between belonging": *Habitus* and the Migration Experience. *The Canadian Geographer* 46(1):63-84.

Masse, W. Bruce

1981 Prehistoric Irrigation Systems in the Salt River Valley, Arizona. *Science* 214:408-415.

Meskell, Lynn

2001 Archaeologies of Identity. In *Archaeology Theory Today*, edited by Ian Hodder, pp. 187-213. Polity Press, Cambridge, U.K..

2002 The Intersections of Identity and Politics in Archaeology. In *Annual Reviews in Anthropology* 31:279-301.

Michelena, R. Louis

1990 Letter by R. Louis Michelena, written in Tempe, Arizona and dated February 3, 1990. MS on file at the Graham County Historical Society, Thatcher, Arizona.

Midvale, Frank

1946 Prehistoric Ruins and Irrigation Features in the Horseshoe Reservoir Region of the Verde River, Central Arizona. Unpublished map on file, Department of Anthropology, Arizona State University, Tempe.

Mills, Barbara J.

1998 Migration and Pueblo IV Community Reorganization in the Silver Creek Area, East-Central Arizona. In *Migration and Reorganization: The Pueblo IV Period in the American Southwest*, edited by Katherine A. Spielmann, pp. 65-80. Anthropological Research Papers No. 51. Arizona State University, Tempe.

2004 Identity, Feasting, and the Archaeology of the Greater Southwest. In *Identity, Feasting, and the Archaeology of the Greater Southwest*, edited by Barbara J. Mills, pp. 1-23. University Press of Colorado, Boulder.

2007 Puebloan Diasporas: Migration, Ethnogenesis, and the Transformation of Western Pueblo Society, A.D. 1000-1400. MS on file, Department of Anthropology, University of Arizona, Tucson.

Mills, Jack P., and Vera M. Mills

1978 *The Curtis Site: A Prehistoric Village in the Safford Valley.* Privately published. Elfrida, Arizona.

Mindeleff, Cosmos

1896 Aboriginal Remains in Verde Valley, Arizona. In *Thirteenth Annual Report of the Bureau of American Ethnology, 1891-1892*, pp. 185-261. Smithsonian Institution, Washington, D.C.

Morrill, R.

1974 *The Spatial Organization of Society.* Duxbury Press, North Scituate.

Motsinger, Thomas N.

1994 An Inside View of Hohokam Architecture. *Kiva* 59(4):395-418.

Nash, R.

1967 *Wilderness and the American Mind.* Yale University Press, New Haven.

Naylor, Thomas H. and Charles W. Polzer, S.J.

1986 *The Presidio and Militia on the Northern Frontier of New Spain, 1570-1700.* The University of Arizona Press, Tucson.

Neely, James A.

1972 Prehistoric Domestic Water Supplies and Irrigation Systems at Monte Alban, Oaxaca, Mexico. Paper presented at the 37th Annual Meeting of the Society for American Archaeology, Miami.

1974 The Prehistoric Lunt and Stove Canyon Sites, Point of Pines, Arizona. Unpublished Ph.D. dissertation, Department of Anthropology, University of Arizona. Tucson. University Microfilms, Inc., Ann Arbor.

1995 Mogollon/Western Pueblo Soil and Water Control Systems of the Reserve Phase: New Data from West-Central New Mexico. In *Soil, Water, Biology, and Belief in Southwestern Prehistoric and Traditional Agriculture*, edited by H. Wolcott Toll, pp. 239-262. Special Publication 2. New Mexico Archaeological Council, Albuquerque.

1997a Foothill Irrigation and Domestic Water Systems of the Safford Valley, Southeastern Arizona. Paper presented to the 62nd Annual Meeting of the Society for American Archaeology, Nashville, Tennessee.

1997b A Developmental Cultural Model for the Safford Valley of Southeastern Arizona and Adjacent Areas. MS prepared for the symposium The Archaeology of a Land Between: Regional Dynamics in the Prehistory and History of Southeast Arizona, October 12-17, 1997, sponsored by the Amerind Foundation,, Dragoon, Arizona.

2001a A Contextual Study of the "Fossilized" Prehispanic Canal Systems of the Tehuacan Valley, Puebla, Mexico. *Antiquity* 75(289):505-506

2001b Prehistoric Agricultural Fields and Water Management Technology of the Safford Valley, Southeastern Arizona. *Antiquity* 75(290):681-682.

2004 Paleoclimatic and Archaeological Contexts. In *The Safford Valley Grids: Prehistoric Cultivation in the Southern Arizona Desert*, edited by William E. Doolittle and James A. Neely, pp. 18 - 30. Anthropological Papers of the University of Arizona, Number 70. The University of Arizona Press, Tucson.

2005a Prehistoric Agriculture and Settlement Systems in Lefthand Canyon, Safford Valley, Southeastern Arizona. In *Inscriptions: Collected Papers in Honor of Richard and Nathalie Woodbury*, edited by R. N. Wiseman, T. O'Laughlin, and C.T. Snow, pp. 145-169. Papers of the Archaeological Society of New Mexico 31, Albuquerque.

2005b Mesoamerican Formative Period Water Management Technology: An Overview with Insights on Development and Regional Interaction. In *New Perspectives on Formative Mesoamerican Cultures*, edited by Terry G. Powis, pp. 127-146. British Archaeological Reports (BAR) International Series 1377. ArchaeoPress, Oxford, U.K.

2008a Initial Observations on Prehistoric Agricultural Strategies in the Safford Basin, Southeastern Arizona. In The Archaeology of a Land Between: Regional Dynamics in the Prehistory and History of Southeastern Arizona, edited by Henry D. Wallace. Archaeological Series, No. 12. The Amerind Foundation, Dragoon, Arizona. Manuscript submitted for publication.

2008b Prehistoric Settlement and Agriculture in Marijilda Canyon, Safford Valley, Southeastern Arizona. (Tentative Title). Unpublished manuscript in possession of the author, Department of Anthropology, University of Texas, Austin.

Neely, James A., and Joseph S. Crary

1998 The Marijilda Canyon Canal: A Complex Irrigation and Domestic Water System in the Safford Valley, Southeastern Arizona. Paper presented at the 63rd Annual Meeting of the Society for American Archaeology, Seattle.

Neely, James A., and William E. Doolittle
2004 Answers and Ideas. In *The Safford Valley Grids: Prehistoric Cultivation in the Southern Arizona Desert*, edited by William E. Doolittle and James A. Neely, pp. 125 - 141. Anthropological Papers of the University of Arizona, Number 70. The University of Arizona Press, Tucson.
2006 Dry-Farming and the Rock-Bordered Grid Fields of the Safford Basin. *Archaeology Southwest* 20(2):7.

Neely, James A., and Carlos A. Rincon Mautner
2004 Los Canales "Fosilizados" del Valle de Tehuacan: Un Proyecto Arqueológico con Levantamiento Cartográfico y Recolección Multidisciplinaria de Muestras. Informe Final al Consejo de Arqueología del Instituto Nacional de Antropologia e Historia de Mexico, Mexico, D.F.

Neely, James A., Carlos A. Rincon Mautner, Raul Hernandez Garciadiego, and Michael J. Aiuvalasit
2005 New Light on the Purrón Dam Complex: A Prehistoric Water Management System in the Tehuacán Valley of Southern México. Paper presented at the 4th Annual meeting of the International Water History Association (IWHA), December 1-4, 2005, Paris, France.

Neily, Robert B., Joseph S. Crary, Gay M. Kinkade, and Stephen Germick
1993 The Owens-Colvin Site Revisited: A Preliminary Report of the Excavations at a Bylas Phase Settlement near Eden, Arizona. Paper presented at the 66th Pecos Conference, Springerville, Arizona.

Neuzil, Anna A.
2001 Ceramics and Social Dynamics: Technological Style and Corrugated Ceramics During the Pueblo III to Pueblo IV Transition, Silver Creek, Arizona. Unpublished M.A. Thesis, Department of Anthropology, University of Arizona, Tucson.
2005a Corrugated Ceramics and Migration in the Pueblo III to Pueblo IV Transition, Silver Creek, Arizona. *Kiva* 71(1):101-124.
2005b In the Aftermath of Migration: Assessing the Social Consequences of Late 13th and 14th Century Population Movements into Southeastern Arizona. Unpublished Ph.D.

dissertation, Department of Anthropology, University of Arizona, Tucson.

2006a Migration in the Safford Basin. *Archaeology Southwest* 20(2):2-3.

2006b Archaeology on the Periphery: Recent Research in the Safford Basin. *Archaeology Southwest* 20(2):1-2.

Neuzil, Anna A. (editor)

2006 Archaeology on the Periphery: Recent Research in the Safford Basin. *Archaeology Southwest* 20(2):whole issue.

Nials, Fred L., James P. Holmlund, and Susan D. Hall

2004 Prehistoric and Historic-Period Canals. In *Ancient Farmers of the Safford Basin: Archaeology of the U.S. 70 Safford-to-Thatcher Project*, edited by Jeffery J. Clark, pp. 59-83. Anthropological Papers 39. Center for Desert Archaeology, Tucson.

Oakes, Yvonne R. and Dorothy A. Zamora

1999 *Archaeology of the Mogollon Highlands: Settlement Systems and Adaptations*. Archaeology Notes 232. Museum of New Mexico, Office of Archaeological Studies, Santa Fe.

Odum, E. P.

1983 *Basic Ecology*. Saunders College Publishing, Philadelphia.

Olmstead, F. H.

1919 Gila River flood control - A report on flood of the Gila River in Graham County, Arizona. U.S. 65th Congress, 3d Session, Doc. 436.

Pagden, Anthony

1990 *Spanish Imperialism and the Political Imagination*. Yale University Press, New Haven.

Pennington, R.

2001 Hunter-gatherer demography. In *Hunter-Gatherers: An Interdisciplinary Perspective*, edited by C. Panter-Brick, R.H. Layton, and P. Rowley-Conway, pp. 170-204. Cambridge University Press, Cambridge, U.K..

Plog, Fred

1978 *An Analytical Approach to Cultural Resource Management: the Little Colorado Planning Unit*. Arizona State University Anthropological Research Papers No. 13 and USDA Forest Service Cultural Resources Report No. 19. Arizona State University, Tempe, and USDA Forest Service, Springerville.

Potter, James M.

1998 The Structure of Open Space in Late Prehistoric Settlements in the Southwest. In *Migration and Reorganization: The Pueblo IV Period in the American Southwest*, edited by Katherine A. Spielmann, pp. 137-163. Anthropological Research Papers No. 51, Arizona State University, Tempe.

Purcell, David E.

2004a *Formative Settlements on the Pinaleño Mountains Bajada: Results of Phased Archaeological Treatment of Sites AZ CC:6:40 and AZ CC:6:43 (ASM) Within the US Highway 191Right-of-Way Between Mileposts 110.40 and 117.60 South of Safford, Graham County, Arizona*. Project Report No. 2001:076. Archaeological Research Services, Inc., Tempe.

2004b Roadrunner Village: Site AZ CC:6:43 (ASM) Feature Descriptions. In *Formative Settlements on the Pinaleño Mountains Bajada: Results of Phased Archaeological Treatment of Sites AZ CC:6:40 and AZ CC:6:43 (ASM) Within the US Highway 191Right-of-Way Between Mileposts 110.40 and 117.60 South of Safford, Graham County, Arizona*, by David E. Purcell, pp. 5.1-102. Project Report No. 2001:076. Archaeological Research Services, Inc., Tempe.

2006a Dynamics of Safford Basin Prehistory: Cultural Affiliation at The Artesia Site and Roadrunner Village, Southeastern Arizona. In *Mostly Mimbres: A Collection of Papers from the 12th Biennial Mogollon Conference,* edited by Marc Thompson, Jason Jurgena, and Lora Jackson, pp. 69-82. El Paso Museum of Archaeology, El Paso.

2006b Eden Phase Occupations at Two Mogollon Villages. *Archaeology Southwest* 20(2):14.

2007 A Cultural Resource Survey of the Salt River Materials Group Buckeye Materials Source Pit, Maricopa County, Arizona. Four Corners Research Report Number 06-294-3. DMG Four Corners Research, Inc., Flagstaff, Arizona. Draft.

Purcell, David E., and Terry L. Coriell

2004a Ground Stone and Shell Tools and Ornaments. In *Formative Settlements on the Pinaleño Mountains Bajada: Results of Phased Archaeological Treatment of Sites AZ CC:6:40 and AZ CC:6:43 (ASM) Within the US Highway 191 Right-of-Way Between Mileposts 110.40 and 117.60 South of Safford, Graham County, Arizona*, by David E. Purcell, pp. 8.1-34. Project Report No. 2001:076. Archaeological Research Services, Inc., Tempe.

2004b The Cultural Identity of Pinaleño Bajada Villagers and Analysis of Their Physical Remains. In *Formative Settlements on the Pinaleño Mountains Bajada: Results of Phased Archaeological Treatment of Sites AZ CC:6:40 and AZ CC:6:43 (ASM) Within the US Highway 191 Right-of-Way Between Mileposts 110.40 and 117.60 South of Safford, Graham County, Arizona*, by David E. Purcell, pp. 11.1-17. Project Report No. 2001:076. Archaeological Research Services, Inc., Tempe.

Quaife, Milo Milton, editor

1935 *Kit Carson's Autobiography.* University of Nebraska Press, Lincoln.

Ramenofsky, Elizabeth L.

1984 *From Charcoal to Banking: The I. E. Solomons of Arizona.* Westernlore Press, Tucson.

Rathje, W. L., W. W. Hughes, D. C. Wilson, M. K. Tani, G. H. Archer, R. G. Hunt, and T. W. Jones

1992 The Archaeology of Contemporary Landfills. *American Antiquity* 57(3):437-447.

Reid, J. Jefferson

1997a Return to Migration, Population Movement, and Ethnic Identity in the American Southwest. In *Vanishing River: Landscapes and Lives of the Lower Verde River*, edited by Stephanie M. Whittlesey, Richard Ciolek-Torrello, and Jeffrey H. Altschul, pp. 629-638. SRI Press, Tucson.

1997b Wickiup 2 at the Grasshopper Spring Site and the Dating of Western Apache Occupation, In *Vanishing River: Landscape and lives of the Lower Verde Valley*, pp. 197-199, edited by Stephanie M. Whittlesey, Richard Cioleck-Torrello, and Jeffrey H. Altschul. SRI Press, Tucson.

Reid, J. Jefferson, and Stephanie M. Whittlesey

1997 *The Archaeology of Ancient Arizona.* The University of Arizona Press, Tucson.

2007 Migration, Population Movement, and Process at Grasshopper Pueblo, Arizona. In *Archaeological Anthropology: Perspectives on Method and Theory*, edited by James M. Skibo, Michael W. Graves, and Miriam T. Stark, pp. 218-235. The University of Arizona Press, Tucson.

Rice, Prudence M.

1987 *Pottery Analysis: A Sourcebook.* The University of Chicago Press, Chicago.

Rinker, Jennifer Rebecca

1998 The Bryce-Smith Project: Irrigated Agriculture and Habitation from A.D. 1000 to 1450, Lefthand Canyon, Safford Valley, Arizona. Unpublished M.A. thesis, Department of Anthropology, University of Texas, Austin.

Rockman, Marcy, and James Steele (editors)

2003 *The Colonization of Unfamiliar Landscapes*. Routledge, London.

Rodgers, James B.

1976 *An Archaeological Investigation of Buckeye Hills East, Maricopa County, Arizona*. Anthropological Research Paper 10. Arizona State University, Tempe.

Rogers, Malcom J.

1939 *Early Lithic Industries of the Lower Basin of the Colorado River and Adjacent Desert Areas*. San Diego Museum Papers No. 3. Museum of Man, San Diego.

1945 An Outline of Yuman Prehistory. *Southwestern Journal of Anthropology* 1:167-198.

1966 *Ancient Hunters of the Far West*. A Copley Book, The Union-Tribune Publishing Company, San Diego.

Rule, Pam

1993 *The Owens-Colvin Site of the Safford Valley*. Museum of Anthropology Publication Number 3. Eastern Arizona College, Thatcher, Arizona.

Russell, Frank

1908 The Pima Indians. *Twenty-Sixth Annual Report of the Bureau of American Ethnology*, 1904-1905, pp. 3-389. Smithsonian Institution, Washington, D.C.

1975 *The Pima Indians*. The University of Arizona Press, Tucson.

Sahlins, Marshall

1972 *Stone-Age Economics*. Aldine Atherton Inc., Chicago and New York.

Sanchez, Bartolomé

1856 Carta del Padre Bartolomé Sanchez, al Padre Prior y Rector Juan Antonio Baltasar en el año 1757. In *Documentos para la Historia México*, pp. 88-94. Cuarta Série, Tomo I. Imprenta de Vincente García Torres, Mexico, D.F.

Sauer, C. D.

1971 *Sixteenth-century North America: The land and the people as seen by the Europeans*. University of California Press, Berkeley.

Sauer, Carl, and Donald Brand

1930 Pueblo Sites in Southeastern Arizona. *University of California Publications in Geography* 3(7): 415–458.

Sayles, E. B.

1945 *The San Simon Branch: Excavations at Cave Creek and in the San Simon Valley, I. Material Culture*. Medallion Paper No. 34. Gila Pueblo, Globe, Arizona.

Schaafsma, Polly

1980 *Indian Rock Art of the Southwest*. School of American Research Press, Santa Fe, and University of New Mexico Press, Albuquerque.

1997 *Rock Art Sites of Chihuahua, Mexico*. Archaeological Note 171. Museum of New Mexico, Office of Archaeological Studies, Santa Fe.

Scarborough, Vernon L.

1993 Water Management in the Southern Maya Lowlands: An Accretive Model for the Engineered Landscape. In *Economic Aspects of Water Management in the Prehispanic New World*, edited by V. L. Scarborough and B. Isaacs, pp. 17-69. Research in Economic Anthropology, Supplement 7. JAI Press, Greenwich.

1994 Maya Water Management. *National Geographic Research and Exploration* 10(2):184-199.

Scarborough, Vernon L., and Gary G. Gallopin

1991 A Water Storage Adaptation in the Maya Lowlands. *Science* 2512:658-662.

Scarborough, Vernon L., Matthew E. Becher, Jeffrey L. Baker, Garry Harris, and Fred Valdez, Jr.

1995 Water and Land at the Ancient Maya Community of La Milpa. *Latin American Antiquity* 6(2):98-119.

Schnapper, Dominique

1994 The Debate on Immigration and the Crisis of National Identity. *West European Politics* 17(2):127-139.

Schwartz, Douglas W.

1970 The Postmigration Culture: A Base for Archaeological Inference. In *Reconstructing Pueblo Societies*, edited by William A. Longacre, pp. 174-193. University of New Mexico Press, Albuquerque.

Seymour, Deni J.

1994 Peripheral Considerations: Defining the Spatial and Physical Correlates of Storage Behavior in Hohokam Structures. *Kiva* 59(4):377-394.

1995 Behavioral Approach to the Recognition of Inconspicuous Apachean Sites. Paper presented at the 68th Pecos Conference, Gila Wilderness, New Mexico.

2002 *Conquest and Concealment: After the El Paso Phase on Fort Bliss*. Prepared for Conservation Division, Directorate of Environment, Fort Bliss, Texas. Report 525/528. Lone Mountain Archaeological Services, El Paso.

2003a The Cerro Rojo Complex: A Unique Indigenous Assemblage In The El Paso Area and Its Implications For The Early Apache. *Proceedings of the XII Jornada Mogollon Conference in 2001*. Geo-Marine, El Paso.

2003b *Protohistoric and Early Historic Temporal Resolution*. Prepared for Conservation Division, Directorate of Environment, Fort Bliss, Texas. Report 560-003. Lone Mountain Archaeological Services, El Paso.

2004a A Ranchería in the Gran Apachería: Evidence of Intercultural Interaction at the Cerro Rojo Site. *Plains Anthropologist* 49(190):153-192.

2004b "Before the Spanish Chronicles: Early Apache in the Southern Southwest." In *Ancient and Historic Lifeways in North America's Rocky Mountains. Proceedings of the 2003 Rocky Mountain Anthropological Conference, Estes Park, Colorado*, edited by Robert H. Brunswig and William B. Butler, pp. 120-142. US Department of the Interior, National Park Service, Estes Park, Colorado.

2005 The Implications of Mobility, Reoccupation, and Low Visibility Phenomena for Chronometric Dating. Ms. on file, Arizona State Museum, Tucson., and <www.seymourharlan.com>.

2006a Current Research in the Study of Protohistoric Mobile Groups: An Introductory Statement. Paper presented at the 2006 Mogollon Conference, Tucson, AZ, October.

2006b *Apache, Spanish, and Protohistoric Archaeology of Fort Bliss*. Historic and Natural Resources Report 03-06, Cultural Resources Program, Directorate of Environment, Fort Bliss, Texas.

2007a Delicate Diplomacy: Seventeenth-Century Sobaípuri Social and Economic Relations in Northwestern New Spain, Part I. *New Mexico Historical Review* 82(4):469-499.

2007b A Syndetic Approach To Identification Of The Historic Mission Site Of San Cayetano Del Tumacácori. *International Journal of Historical Archaeology* 11(30): 269-296.

2007c An Archaeological Perspective on the Hohokam-Pima Continuum. *Old Pueblo Archaeology Bulletin.* 51(December):1-7.

2007d Geronimo's Wickiup: Context, Contingency, and Mobile Group Hut Signatures. Ms. on file, Arizona State Museum, Tucson, and <www.seymourharlan.com>.

2008a Evaluating Eyewitness Accounts of Native Peoples Along The Coronado Trail From The International Border To Cibola. *New Mexico Historical Review.* In press.

2008b Surfing Behind The Wave: A Counterpoint Discussion Relating To "A Ranchería In The Gran Apachería." *Plains Anthropologist.* In press.

2008c Delicate Diplomacy: Seventeenth-Century Sobaípuri Social and Economic Relations in Northwestern New Spain, Part II. *New Mexico Historical Review.* In press.

2008d Apache Plain and Other Plainwares on Apache Sites in the Extreme Southern Southwest. In *Chasing Chaco & The Southwest: Papers in Honor of Frances Joan Mathien*, edited by Regge N. Wiseman, Thomas C. O'Laughlin, Cordelia T. Snow and Cathy Travis, pp. 163-186. The Archaeological Society of New Mexico: 34.

2008e The Canutillo Complex: Evidence of Protohistoric Mobile Occupants in the Southern Southwest. *Kiva.* In press.

Seymour, Deni J. and George Robertson

2008 A Pledge of Peace: Evidence of the Cochise-Howard Treaty Campsite. *Historical Archaeology* 42(4).

Seymour, Gregory R., Richard V. N. Ahlstrom, and David P. Doak (editors)

1997 *The Sanchez Copper Project: Vol. 1. Archaeological Investigations in the Safford Valley, Graham County, Arizona.* Archaeological Report 94-82 (Revised 1997). SWCA, Inc., Environmental Consultants, Tucson.

Shafer, Harry J.

1999 The Mimbres Classic and Postclassic: A Case for Discontinuity. In *The Casas Grandes World*, edited by Curtis R. Schaafsma and Carroll L. Riley, pp 121-133. University of Utah Press, Salt Lake City.

Slaughter, Mark C., and Heidi Roberts (editors)

1996 *Excavation of the Gibbon Springs Site: A Classic Period Village in the Northeastern Tucson Basin.* Report No. 94-87. SWCA, Inc., Environmental Consultants, Tucson.

Slifer, Dennis
2000 *The Serpent and the Sacred Fire: Fertility Images in Southwestern Rock Art.* Museum of New Mexico Press, Santa Fe.

Smith, Alexa M., David R. Abbott, and Anne Cherryman
2004 A Thousand Years in the Making: Prehistoric Ceramics at Roadrunner Village and The Artesia Site. In *Formative Settlements on The Pinaleño Mountains Bajada: Results of Phased Archaeological Treatment of Sites AZ CC:6:40 and A CC:6:43 (ASM) Within the US Highway 191 Right-of-way Between Mileposts 110.40 and 117.60 South of Safford, Graham County, Arizona*, by David. E. Purcell, pp. 9.1-9.59. Project Report No. 2001:076. Archaeological Research Services, Inc., Tempe.

Smith, Michael E. and Francis F. Berdan (editors)
2003a *The Postclassic Mesoamerican World.* University of Utah Press, Salt Lake City.

Smith, Susan J.
2004 Pollen Analysis. In *Ancient Farmers of the Safford Basin: Archaeology of the U.S. 70 Safford-to-Thatcher Project*, edited by Jeffrey J. Clark, pp. 165-171. Anthropological Papers 39. Center for Desert Archaeology, Tucson.

Snow, Dean R.
1995 Migration in Prehistory: The Northern Iroquoian Case. *American Antiquity* 60(1):59-80.
1996 More on Migration in Prehistory: Accommodating New Evidence in the Northern Iroquoian Case. *American Antiquity* 61(4):791-796.

Southworth, C. H.
1919 The History of Irrigation Along the Gila River. In Hearings Before the Committee on Indian Affairs, House of Representatives, Sixty-sixth Congress. Government Printing Office, Washington, D.C.

Spicer, Edward H.
1962 *Cycles of Conquest: The Impact of Spain, Mexico, and the United States on the Indians of the Southwest, 1533-1960.* The University of Arizona Press, Tucson.

Stanbury, Pamela C.
1996 The Utility of Tradition in Sri Lankan Bureaucratic Irrigation: The Case of the Kirindi Oya Project. In *Canals and Communities: Small-Scale Irrigation Systems*, edited by

Jonathan B. Mabry, pp. 210-226. The University of Arizona Press, Tucson.

Stockel, H. Henrietta

2007 Rocks, Waters, Earth: Chiricahua Apache Spiritual Geography. *Journal of the West*. 46(4).

Stone, Connie L.

1991 *The Linear Oasis: Managing Cultural Resources Along the Lower Colorado River.* Cultural Resources Series No. 6. USDI Bureau of Land Management, Arizona State Office, Phoenix.

Stone, Tammy

2003 Social Identity and Ethnic Interaction in the Western Pueblos of the American Southwest. *Journal of Archaeological Method and Theory* 10(1):31-67.

Sweeney, Edwin R.

1991 *Cochise: Chiricahua Apache Chief.* University of Oklahoma Press, Norman.

1992 *Merejildo Grijalva, Apache Captive, Army Scout.* Southwestern Study Series No. 96. University of Texas, El Paso.

Talley, George Thomas

1970 History of the Graham Canal. MS on file at the Graham County Historical Society, Safford, Arizona. (Date is approximate).

Taylor, C.

1983 AZ CC:2:31 (BLM) site survey record. USDI Bureau of Land Management, Safford Office, Safford.

Teague, Lynn S.

1993 Prehistory and the Traditions of the O'odham and Hopi. *Kiva* 58(4):435-454.

Thiel, J. Homer, and Jonathan B. Mabry (editors)

2006 *Rio Nuevo Archaeology Program, 2000-2003: Investigations at the San Agustin Mission and Mission Gardens, Tucson Presidio, Tucson Pressed Brick Company, and Clearwater Site.* Technical Report No. 2004-11. Center for Desert Archaeology, Tucson.

Torres-Saillant, Silvio

1997 Diaspora and National Identity: Dominican Migration in the Postmodern Society. *Migration World Magazine* 25(3):18-22.

Tuohy, Donald R.

1960 Archaeological Survey and Excavation in the Gila River Channel Between Earven Dam Site and Buttes Reservoir Site, Arizona. Prepared for the National Park Service. MS on file, Arizona State Museum Library, Tucson.

Tyberg, Joel Jay

2000 Influences, Occupation, and Salado Development at the Solomonsville Site. Unpublished M.A. thesis, University of Colorado, Boulder.

Upton, Dell
1996 Ethnicity, Authenticity, and Invented Traditions. *Historical Archaeology* 30(2):1-7.

U. S. Department of the Interior
1935 Aerial Photographic Mosaic, Gila-San Pedro Watersheds, New Mexico and Arizona. Project #3250, Scale of 1 Inch = 2640 Feet. Fairchild Aerial Surveys, Inc. for the Soil Erosion Service, U. S. Department of the Interior, Los Angeles.

Van Keuren, Scott
1999 *Ceramic Design Structure and the Organization of Cibola White Ware Production in the Grasshopper Region, Arizona.* Arizona State Museum Archaeological Series, Number 191. The University of Arizona Press, Tucson.

Van West, Carla, and Jeffrey H. Altschul
1997 Environmental Variability and Agricultural Economics along the Lower Verde River, A.D. 750-1450. In *Vanishing River: Landscapes and Lives of the Lower Verde Valley: The Lower Verde Archaeological Project*, edited by Stephanie. M. Whittlesey, Richard Ciolek-Torrello, and Jeffrey H. Altschul, pp. 337-392. SRI Press, Tucson.

Van West, Carla R., and Jeffrey S. Dean
2000 Environmental Characteristics of the A.D. 900-1300 Period in the Central Mesa Verde Region. *Kiva* 66(1):19-44.

Vertovec, Steven
2001 Transnationalism and identity. *Journal of Ethnic and Migration Studies* 27(4):573-582.

Wallace, Henry D., and James P. Holmlund
1986 *Petroglyphs of the Picacho Mountains, South–Central Arizona.* Anthropological Papers No. 6. Institute for American Research, Tucson.

Wasley, William W.
1962 A Ceremonial Cave on Bonita Creek, Arizona. *American Antiquity* 27(3):380-394.

Wauchope, Robert (editor)
1956 Seminars in Archaeology: 1955. *American Antiquity* 22(2): Part 2.

Welch, John R.

1995 Preservation, Research, and Public Interpretation at Pueblo Devol, an Arizona Cliff Dwelling. *Kiva* 61(2):121-143.

Whalen, Michael E. and Paul E. Minnis

2001 *Casas Grandes and Its Hinterland: Prehistoric Regional Organization in Northwest Mexico*. The University of Arizona Press, Tucson.

Wheat, Joe Ben

1955 *Mogollon Culture Prior to A.D. 1000*. American Anthropological Association Memoir No. 82. Menasha, Wisconsin.

White, William B.

1994 Cast Shadows, A Lizard's Tail, and Prehistoric Time Reckoning: A Calendrical Petroglyph Along the Lower Colorado River. In *Recent Research Along the Lower Colorado River*, edited by Joseph. A. Ezzo, pp. 69–80. Technical Series No. 51. Statistical Research , Inc., Tucson.

Whittlesey, Stephanie M

1995 Mogollon, Hohokam, and O'otam: Rethinking the Early Formative Period in Southern Arizona. *Kiva* 60(4):465-480.

Whittlesey, Stephanie M., Richard S. Ciolek-Torrello, and J. Jefferson Reid

2000 Salado: The View from the Arizona Mountains. In *Salado*, edited by Jeffrey S. Dean, pp. 241-262. Amerind Foundation New World Studies Series Number Four. Amerind Foundation and University of New Mexico Press, Albuquerque.

Whittlesey, Stephanie M., and Robert A. Heckman

2000 Culture History and Research Background. In Concluding Thoughts. In *Prehistoric Painted Pottery of Southeastern Arizona*, by Robert A. Heckman, Barbara K. Montgomery, and Stephanie M. Whittlesey, pp. 117-128. Technical Series 77. Statistical Research, Inc., Tucson.

Wilcox, David R.

1981 The Entry of Athapaskans into the American Southwest: the Problem Today. In *The Protohistoric Period in the North American Southwest. AD 1450–1700*, edited by David R. Wilcox and William B. Masse, pp. 275-281. Anthropological Research Papers No. 24. Arizona State University, Tempe.

1991 Hohokam Religion: An Archaeologist's Perspective. In *The Hohokam: Ancient People of the Desert*, edited by David. G. Noble, pp. 47-61. School of American Research Press, Santa Fe.

Wilkinson, Tony J.

2003 *Archaeological Landscapes of the Near East.* The University of Arizona Press, Tucson.

Williams, Oran A.

1937 Settlement and Growth of the Gila Valley in Graham County as a Mormon Colony 1879-1900. Unpublished M.A. thesis, Department of History, University of Arizona, Tucson.

Wilshusen, Richard H., and Scott G. Ortman

1999 Rethinking the Pueblo I Period in the San Juan Drainage: Aggregation, Migration, and Cultural Diversity. *Kiva* 64(3):369-400.

Winship, George Parker (editor and translator)

1990 *The Journey of Coronado,* by Pedro de Castañeda. Dover Publications, Inc., New York.

Woodbury, Richard B., and James A. Neely

1972 Water Control Systems of the Tehuacán Valley. In *Chronology and Irrigation. The Prehistory of the Tehuacán Valley, Vol. 4*, edited by R. S. MacNeish, pp. 81-153. University of Texas Press for the R. S. Peabody Foundation, Austin.

Woodson, M. Kyle

1995 The Goat Hill Site: A Western Anasazi Pueblo in the Safford Valley of Southeastern Arizona. Unpublished M.A. Thesis, Department of Anthropology, University of Texas, Austin.

1999 Migrations in Late Anasazi Prehistory: The Evidence from the Goat Hill Site. *Kiva* 65(1):63-84.

2006 The Goat Hill Site and Ancestral Pueblo Migrations into the Safford Basin. *Archaeology Southwest* 20(2):4.

Wright, Thomas E.
2001 *A Cultural Resources Survey Along U.S. Highway 191 Between Mileposts 110.40 and 117.60, South of Safford, Graham County, Arizona.* Project Report No. 2000:079. Archaeological Research Services, Inc., Tempe.

Young, David A., and Robert L. Bettinger
1992 The Numic Spread: A Computer Simulation. *American Antiquity* 57(1):85-98.

Zedeño, Maria Nieves, Richard Stoffle, Genevieve Dewy–Hefley, and David Shaul
1994 *Storied Rocks: American Indian Inventory and Interpretation of Rock Art on the Nevada Test Site.* Technical Report No. 93. Prepared by the Bureau of Applied Research in Anthropology, University of Arizona, Tucson. Desert Research Institute, University of Nevada, Las Vegas.

# CONTRIBUTORS

**Jeffery J. Clark**, Ph.D. - Preservation Archaeologist, Center for Desert Archaeology, Tucson, Arizona.

Publications:

2004 *Ancient Farmers of the Safford Basin* [editor]. Anthropological Papers 39. Center for Desert Archaeology, Tucson.

2004 Prehistoric Demography in the Southwest: Migration, Coalescence, and Hohokam Population Decline [second author, with Brett J. Hill, William H. Doelle, and Patrick D. Lyons]. *American Antiquity* 69(4):689-716.

2004 Tracking Cultural Affiliation: Enculturation and Ethnicity. In *Identity, Feasting, and the Archaeology of the Greater Southwest*, edited by Barbara J. Mills, pp. 42-73. University of Colorado Press, Boulder.

2001 *Tracking Prehistoric Migrations: Pueblo Settlers Among the Tonto Basin Hohokam*. Anthropological Papers of the University of Arizona No. 65. The University of Arizona Press, Tucson.

1997 Patterns in the Pathways [second author, with Sarah Herr]. *Kiva* 62(4):365-389.

**Joseph A. Ezzo**, Ph.D. - Principal Investigator, SWCA Environmental Consultants, Tucson, Arizona.

Publications:

2006 The Earth, the Sky: A Functional Comparison of Surface Archaeology in the Lower Colorado River Region and the Nazca Plain. In *Beginnings: Proceedings of the 2005 Three Corners Conference*, edited by C. Slaughter, Gregory R. Seymour, and Laureen M. Perry, pp. 101-111. The Three Corners Conference, distributed by Nevada Archaeological Association, Reno.

2003 Fish, Flesh, or Fowl: in Pursuit of a Diet-Mobility-Climate Continuum Model in the Cis-Baikal. [first author, with Andrzej Witold Weber, Olga I. Goriunova, and Vladimir I. Bazaliiskii] In *Northern Hunter Gatherers: Research Studies Volume 1.*

*Prehistoric Foragers of the Cis-Baikal: Proceedings from the First Conference of the Baikal Archaeology Project*, edited by A. W. Weber and H. G. MacKenzie, pp. 123-131. Canadian Circumpolar Institute Press, Edmonton.

2002 Migration, Residential Reorganization, and Spatial Group Composition at Grasshopper Pueblo, Arizona. [first author, with T. Douglas Price] *Journal of Archaeological Science* 29:499-520.

2000 A Late Archaic-period Domestic Dog Burial from the Tucson Basin, Arizona. [first author, with Mary C. Stiner] *Kiva* 66:291-305.

1999 A Heterarchical Perspective on Aggregated Pueblo Social Organization: A Case Study of the Grasshopper Ruin, East-central Arizona. In *Sixty Years of Mogollon Archaeology: Proceedings of the Ninth Mogollon Conference*, edited by S. M. Whittlesey, pp. 17-24. Statistical Research Press, Tucson.

**Stephen H. Lekson**, Ph.D. - Curator and Professor of Anthropology, University of Colorado Museum of Natural History, University of Colorado, Boulder, Colorado.

Publications:

2007 *The Architecture of Chaco Canyon* [editor]. University of Utah Press, Salt Lake City.

2006 *Archaeology of the Mimbres Region, Southwestern New Mexico, U.S.A.* British Archaeological Reports, International Series 1466, Oxford.

2006 *The Archaeology of Chaco Canyon* [editor]. School of American Research Press, Santa Fe.

2002 *Salado Archaeology of the Upper Gila, New Mexico*. University of Arizona Press, Tucson.

1999 *Chaco Meridian: Centers of Political Power in the Ancient Southwest.* Altamira Press, Walnut Creek, California.

**Everett J. Murphy** - Arizona Site Steward, Safford, Arizona.

No prior publications.

**James A. Neely**, Ph.D. - Professor Emeritus, Department of Anthropology, University of Texas, Austin, Texas.

Publications:

2005 Prehistoric Agricultural and Settlement Systems in Lefthand Canyon, Safford Valley, Southeastern Arizona. In *Inscriptions: Papers in Honor of Richard and Nathalie Woodbury*, edited by R. N. Wiseman, T. O'Laughlin, and C. T. Snow, pp. 145-169. Papers of the Archaeological Society of New Mexico 31, Albuquerque.

2005 Mesoamerican Formative Period Water Management Technology: An Overview with Insights on Development and Regional Interaction. In *New Perspectives on Formative Mesoamerican Cultures*, edited by Terry G. Powis, pp. 127-146. British Archaeological Reports (BAR) International Series 1377. Archaeo Press, Oxford.

2004 *The Safford Valley Grids: Prehistoric Cultivation in the Southern Arizona Desert*. [second editor, with William E. Doolittle] Anthropological Papers of the University of Arizona 70. University of Arizona Press, Tucson.

2003 Sassanian and Early Islamic Water Control and Irrigation Systems on the Deh Luran Plain, Southwestern Iran. In "Production and the Exploitation of Resources", edited by M. G. Morony, pp. 291-310. *The Formation of the Classical Islamic World, Vol. 11*. Ashgate Publishing Limited, Aldershot, U.K.

1999 The Gristmill at Mission San José y San Miguel de Aguayo: Insights into the Technology and Agricultural Focus of Spanish Colonial Texas. *Bulletin of the Texas Archaeological Society* 70:215- 239

**Anna A. Neuzil**, Ph.D. - Ceramic Analyst, Tierra Right of Way Services Ltd., Tucson, Arizona.

Publications:

2006 *Ten Sites in the Gila Box Riparian National Conservation Area, Northeast of Safford, Graham County, Arizona*. Technical Report No. 2006-106. Center for Desert Archaeology, Tucson.

2006 *A Cultural Resources Survey of 1075 Acres in the Aravaipa Canyon Wilderness Area, Pinal County, Arizona*. Technical Report No. 2006-104. Center for Desert Archaeology, Tucson.

2006 *An Analysis of Whole Vessels from the Mills Collection curated at Eastern Arizona College, Thatcher, Arizona* [first author, with Patrick D. Lyons]. Technical Report No. 2005-001. Center for Desert Archaeology, Tucson.

2005 Corrugated Ceramics and Migration in the Pueblo III to Pueblo IV Transition, Silver Creek, Arizona. *Kiva* 71(1):101-124.

2004 Chapter 14: Safford Basin Past and Present [second author, with Jeffery J. Clark and Patrick D. Lyons]. In *The Ancient Farmers of the Safford Basin: Archaeology of the U.S. 70 Safford-to-Thatcher Project*, edited by Jeffery J. Clark, pp. 207-233. Anthropological Papers No. 39, Center for Desert Archaeology, Tucson.

**David E. Purcell**, M.A. - Archaeologist, Logan Simpson Design, Inc., Tempe, Arizona.

Publications:

2006 Dynamics of Safford Basin Prehistory: Cultural Affiliation at The Artesia Site and Roadrunner Village, Southeastern Arizona. In *Mostly Mimbres: A Collection of Papers from the 12th Biennial Mogollon Conference,* edited by Marc Thompson, Jason Jurgena, and Lora Jackson, pp. 69-82. El Paso Museum of Archaeology, El Paso.

2006 Archaeological Investigations Of Xiongnu Sites in the Tamir River Valley: Results of the 2005 Joint American-Mongolian Expedition to *Tamiryn Ulaan Khoshuu*, Ogii nuur, Arkhangai Aimag, Mongolia [first author, with Kimberly C. Spurr]. *The Silk Road* 4(1):20-32.

2006 Big Bend Quarry Revisited: Preliminary Results of the 9,400 Acre Laughlin Ranch Project in Bullhead City, Arizona [first author, with Adam Okun and David H. Greenwald]. In *Beginnings: Proceedings of the 2005 Three Corners Conference*, edited by Mark C. Slaughter, Gregory R. Seymour, and Laureen M. Perry, pp. 79-99. The Three Corners Conference, distributed by Nevada Archaeological Association, Reno.

2005 Jornada Mogollon Utilization of White Sands National Monument [first author, with David H. Greenwald and Diane E. White]. In *Archaeology Between the Borders: Papers from the 13th Biennial Jornada Mogollon Conference*, edited by Marc Thompson, Jason Jurgena, and Lora Jackson, pp. 95-114. El Paso Museum of Archaeology, El Paso.

2000 Peach Springs Revisited: Surface Recording and Excavation on the South Chaco Slope, New Mexico [second author, with Dennis Gilpin]. In *Great House Communities Across the Chacoan Landscape*, edited by John Kantner and Nancy M. Mahoney, pp.

28-38. Anthropological Papers Number 64. The University of Arizona Press, Tucson.

**Deni J. Seymour**, Ph.D. - Adjunct Researcher, University of Colorado Museum, Boulder, Colorado.

Publications:

2008 A Pledge of Peace: Evidence of the Cochise-Howard Treaty Campsite [first author, with George Robertson]. *Historical Archaeology* 42(4).

2007 A Syndetic Approach To Identification Of The Historic Mission Site Of San Cayetano Del Tumacácori. *International Journal of Historical Archaeology* 11(3): 269-296.

2007 Delicate Diplomacy on a Restless Frontier: Seventeenth-Century Sobaípuri Social and Economic Relations in Northwestern New Spain, Part 1. *New Mexico Historical Review* 82(4): 469-499.

2004 A Ranchería in the Gran Apachería: Evidence of Intercultural Interaction at the Cerro Rojo Site. *Plains Anthropologist* 49(190):153-192.

1997 Finding History in the Archaeological Record; The Upper Piman Settlement of Guevavi. *Kiva* 62(3):245-260.

**M. Kyle Woodson**, M.A. - Project Director, Cultural Resource Management Program, Gila River Indian Community, Arizona, and a Ph.D. candidate in the Anthropology program at the School of Human Evolution and Social Change, Arizona State University.

Publications:

2008 The Safford Basin and Aravaipa Creek: A Cultural Melting Pot of the Ancient Past [second author, with Anna A. Neuzil]. In *The Archaeology of a Land Between: Regional Dynamics in the Prehistory and History of Southeastern Arizona*, edited by Henry D. Wallace. Amerind Foundation New World Studies Series. In press: manuscript submitted 2006.

2008 The Growth and Consolidation of Settlement Complexes and Irrigation Communities on the Middle Gila [third author, with John C. Ravesloot, and Glen E. Rice]. In *Hohokam Trajectories in World Perspective*, edited by Paul R. Fish and Suzanne K. Fish. Amerind Foundation New World Studies Series. In press: manuscript submitted 2006.

2007 Hohokam Pottery Production Areas and the Organization of

Ceramic Production and Exchange in the Phoenix Basin. *Kiva*. In press: manuscript submitted 2004.

1999 Migrations in Late Anasazi Prehistory: The Evidence from the Goat Hill Site. *Kiva* 65:63–84.

1999 Exploring Late-Prehistoric Mortuary Patterns in Southeastern Arizona and Southwestern New Mexico [first author, with Thomas E. Jones and Joseph S. Crary]. In *Sixty Years of Mogollon Archaeology: Papers from the Ninth Mogollon Conference, Silver City, New Mexico, 1996*, edited by Stephanie M. Whittlesey, pp. 67–79. SRI Press, Tucson.